Bartłomiej Belcarz

Colour illustrations by
Artur Juszczak

Dewoitine D.520

STRATUS

Published in Poland in 2005 by STRATUS
Artur Juszczak, Po. Box 123, 27-600 Sandomierz 1, Poland
e-mail: arturj@mmpbooks.biz
for
Mushroom Model Publications,
36 Ver Road, Redbourn,
AL3 7PE, UK.
e-mail: rogerw@mmpbooks.biz

ISBN 83-89450-09-7

Editor in chief Roger Wallsgrove

Editorial Team Bartłomiej Belcarz
Robert Pęczkowski
Artur Juszczak

Colour Drawings Artur Juszczak
Scale plans Marek Ryś
Photo Editor Dariusz Karnas

Get in the picture!

Do you have photographs of historical aircraft, airfields in action, or original and unusual stories to tell? MMP would like to hear from you! We welcome previously unpublished material that will help to make MMP books the best of their kind. We will return original photos to you and provide full credit for your images. Contact us before sending us any valuable material: ***rogerw@mmpbooks.biz***

Acknowledgements: Author would like to thanks following individuals: Many Souffan, Serge Joanne, Krasimir Aleksandrov, Tomasz Kopański, Przemysław Skulski, Jiri Rajlich, James Crow, James Kightly, Thomas Genth, Pierre-Clement Got.

Printed by: Drukarnia Diecezjalna, ul. Żeromskiego 4, 27-600 Sandomierz
tel. (15) 832 31 92; fax (15) 832 77 87
www.wds.pl marketing@wds.pl
PRINTED IN POLAND

Table of contents

Bulgarian D.520 at Karlovo airfield.
K. Aleksandrov coll.

Development

A government programme of air force modernisation was announced in France in early 1937. Initially, the Ministere de l'Air showed no interest in the Dewoitine D.520. As the type was not among those selected for the prototype stage, it was then developed by the company as a private venture.

Construction of the prototype used the experience obtained when developing the Dewoitine D 513. The team led by Robert Costello and Jacques Henrat designed an aeroplane able to compete with the Morane 406 or Bloch 151. Although critics reproached the designers for using a single-spar wing which was apparently inappropriate, tests did not confirm these reservations. A mock-up of the aeroplane was built in mid-1938, and on 2 October the prototype, designated D.520-01, was test flown. The aeroplane was powered by the Hispano Suiza HS 12Y-29 engine of 655 kW (890 hp). After a few flights, the prototype was damaged by Marcel Doret on 28 November 1938, when the pilot forgot to lower the undercarriage during landing. Meanwhile, work continued on the second prototype, which included a number of changes that resulted from the first tests. The wing was redesigned, the radiator being repositioned under the fuselage. The second prototype was first flown on 6 January 1939. Already the initial trials gave outstanding results. As a result a government delegation paid a visit to get acquainted with the new design. Test flying by Marcel Detroyat confirmed the excellent performance of the aeroplane. The D.520 was thus earmarked for state trials that commenced

D.520.01 prototype
M. Souffan coll.

at Villacoublay (near Paris) in late January 1939. During these C. Rozanoff achieved a speed of 520 km/h, and even 825 km/h in a dive (7 February 1939). Once the trials were completed, the aeroplane returned to the factory at Toulouse, where a more powerful engine was fitted (Hispano Suiza HS 12Y-31 – 803 kW/1100 hp).

The aeroplane in this configuration achieved a speed of 550 km/h, even though it did not have the supercharger. This was fitted in the third prototype, D.520-03, first flown by Marcel Doret on 5 May 1939. The aeroplane then took part in a series of state trials at Villacoublay and Cazaux, completed on 12 May 1940. These had little influence on the order placed by the Air Ministry.

The first contract was signed on 14 March 1939 and called for 200 D.520s. In June of that year, under the new air force equipment plan, the D.520 order was increased by 600. Eventually, after the war broke out, the number of D.520s on order reached a total of 1,280. In January 1940 the Aeronavale (naval aviation) placed an option for another 120 D.520s.

The first production D.520 was flown on 31 October 1939. It had a Hispano Suiza HS 12Y-45 engine. Achieving the proper rate of manufacture was difficult due to problems with engine deliveries. Therefore it was proposed to fit the D.520 with the Rolls-Royce Merlin III or the Allison V-1710-C5. Eventually these plans were abandoned, though not before a Merlin engine had been tested in a prototype.

By the time the operations in the West commenced (10 May 1940) a total of 228 D.520s had been built, of which 75 were delivered to combat units. Most of the remaining ones were undergoing modification at Toulouse-Blagnac.

Signing the German-French armistice did not terminate D.520 production. Newly built D.520s were introduced into service with the Vichy air force, phasing out the Morane 406, Bloch 152 and Curtiss Hawk H 75 fighters. Production plans called for 550 D.520 aircraft to be built by mid-1943. These plans had not been carried out by the time the Germans entered Southern France.

In January 1943 D.520 production was resumed, and overhauls and repairs of nearly 250 D.520s taken over by the Luftwaffe were also ordered. These aircraft were used in training, some also found their way to Italy and Bulgaria

Final assembly of the D.520. Dewoitine factory at Toulouse-Francazal.
M. Souffan coll.

Prototypes and projects

D.520-01

The first prototype, powered by the Hispano-Suiza HS 12Y-21 engine driving a wooden, two-blade, fixed pitch propeller. Short fin in racing aircraft style. Two radiators under the wings. The aeroplane was unarmed and was fitted with a tail skid. During trials the engine was replaced with the Hispano-Suiza HS 12Y-29 driving a three-blade propeller, and a single under-fuselage radiator was fitted. This enabled it to reach a speed of 520 km/h, and cured the engine overheating problem.

D.520-01, prototype. 1/72 scale.

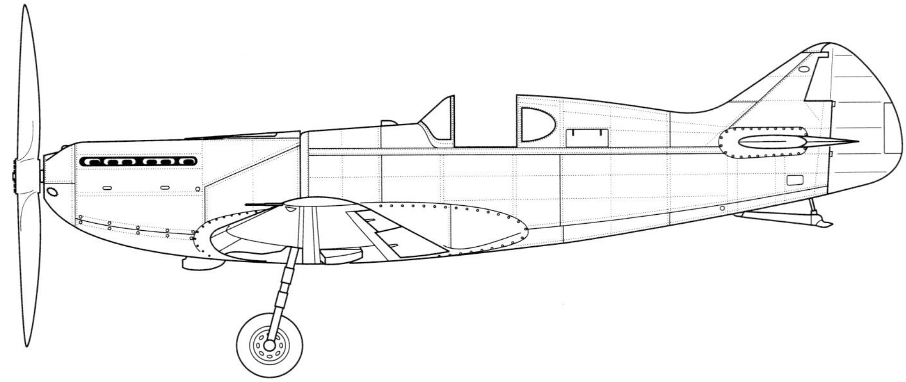

D.520-02

The second prototype, powered by the Hispano-Suiza HS 12Y-29. Fitted with a central radiator, similar to the first prototype, an enlarged fin, strengthened undercarriage, and a new cockpit canopy. The aeroplane was armed with a 20 mm cannon firing through propeller disc, and two under-wing 7.5 mm machine guns.

D.520-02, prototype. 1/72 scale.

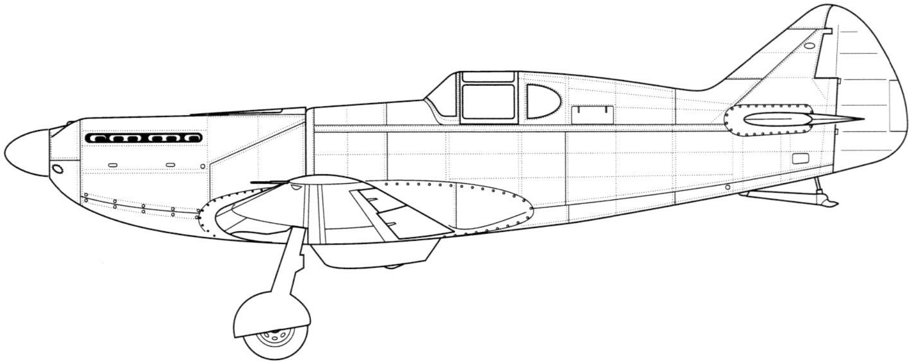

D.520-03

The third prototype was similar to the second, but it was powered by the Hispano-Suiza HS 12Y-31 engine with the Szydlowski supercharger. A tail wheel was introduced.

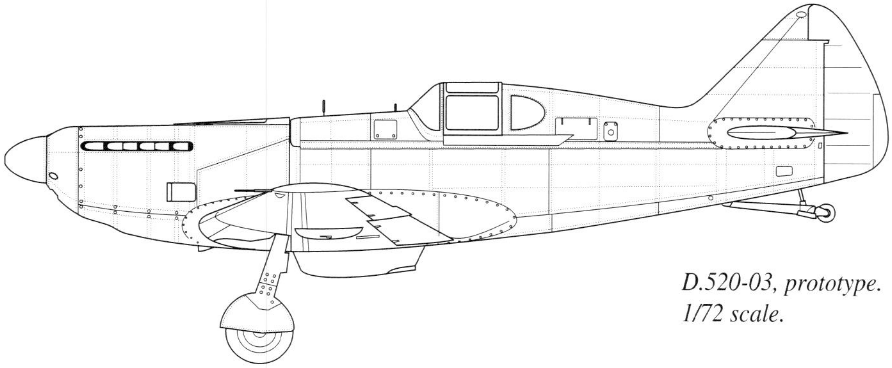

D.520-03, prototype.
1/72 scale.

D.521-01

Expecting delivery problems with the HS 12Y 45 engine, this aeroplane was modified for the new Rolls-Royce Merlin III, planned to be used from aeroplane no. 251 on. One of production aircraft was fitted with the new engine, and two 20 mm cannon plus two machine guns. The heavier engine resulted in centre of gravity problems. The aeroplane was first flown on 9 February 1940. Due to problems the trials were halted on 23 March 1940 after seven flights. The test machine (no. 41) was reconverted to its original form and handed over for further testing.

D.521
1/72 scale.

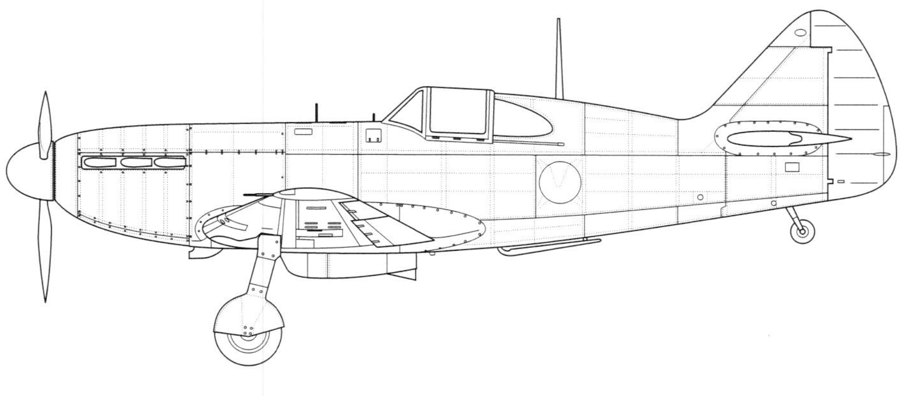

D.522

Planned development version. Similar to the D.521-01, it was developed due to problems with deliveries of H-S engines. It was planned to use the American Allison V-1710C-1 engine. The project was abandoned after the armistice (22 June 1940).

D.523-45

Development version. New Hispano-Suiza HS 12Y-51 engine with the Szydlowski-Planiol supercharger was used, allowing it to develop 1,000 hp (735 kW). It was planned to fit the new engine from aircraft no. 601 onwards. The first aeroplane with the new engine was first flown on 9 May 1939, achieving a top speed of 570 km/h.

D.524

Development version. It was planned to use the new 1,200 hp Hispano-Suiza HS 12Y-89ter engine, to achieve a top speed of 616 km/h at an altitude of 7,000 m. Work on fitting the engine in aeroplane no. 41 (D.521) commenced, but it was never flown.

D.525

Development version of the D.523 with a new Hispano-Suiza supercharger. It was planned to start production from aeroplane no. 751 onwards.

D.530

Planned development version. It was planned to fit a 1,400 hp Rolls-Royce Merlin engine or a 1,800 hp Hispano-Suiza 12Y. The prototype was never built.

D.520 was the only French fighter which was able to match German ones during the French Campaign in 1940.

P. Skulski coll.

D.520 Z

Planned development version of the D.520 with a modified and lengthened engine cowling, 882 kW (1,200 hp) Hispano-Suiza HS 12Z engine, and a new cooling system. Also the undercarriage was redesigned, and the armament would consist of three 20 mm cannon. Calculated top speed of 650 km/h. Preliminary specification for the project was prepared in 1939. It was not until 1942 that Vichy applied to the Germans for a permission to start construction of three prototypes and 230 production aircraft. Permission was granted for one prototype and 105 production aircraft. In March 1943, after the Germans occupied Southern France, construction of the prototype was terminated.

D.520Z 1/72 scale.

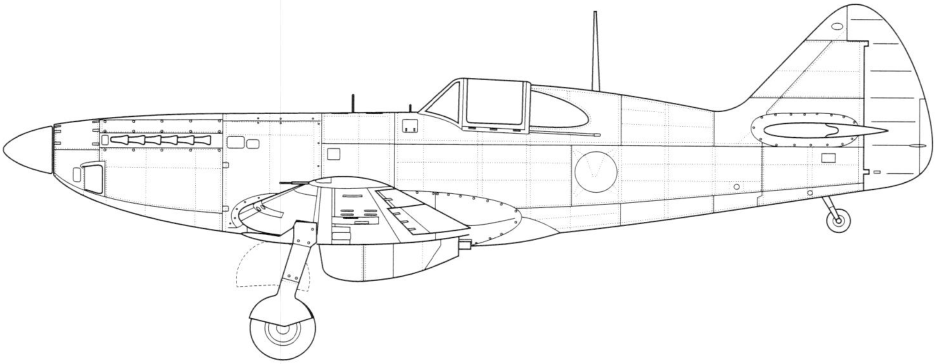

D.550

Racer version of the D.520 with a Hispano-Suiza HS 12Ycrs engine and smaller span wings. The aeroplane with the HS 12Ycrs engine was first flown on 23 June 1939 at Toulouse-Francazal. In November, with the new HS 12Y51 engine it achieved a top speed of 702 km/h. The project was built as a private venture by E. Dewoitine. On 24 February 1940 the aeroplane achieved a speed of 703 km/h. The test was interrupted due to a rise of the engine temperature before it reached its full power. The last flight took place on 18 May 1940.

D.550 1/72 scale.

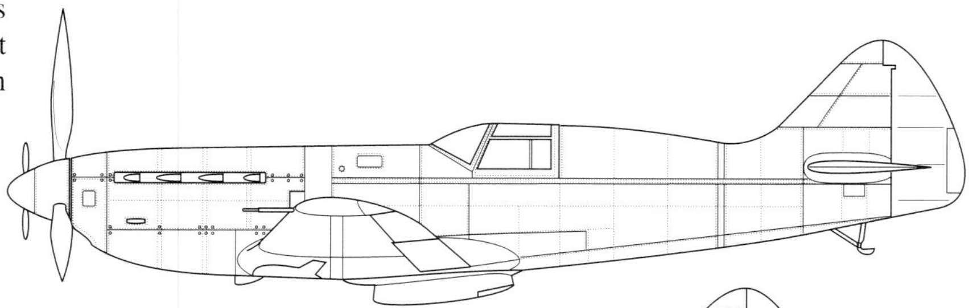

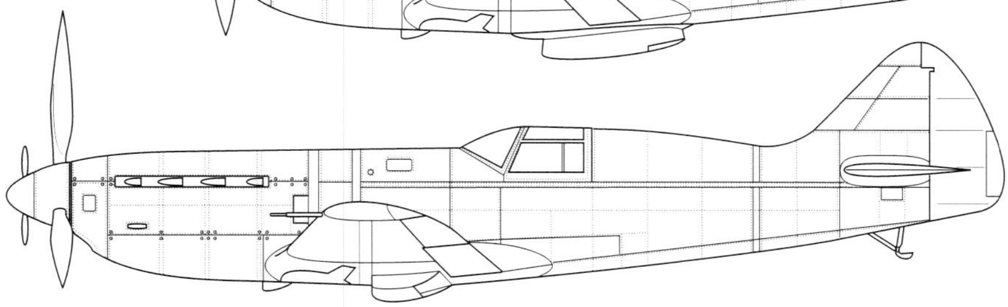

D.551/D.552

Military version of the D.550 with the Hispano-Suiza HS 12Y 51 engine. 18 aircraft were at various stages of completion at the time of the armistice, but none were completed. On 9 August 1940 the Germans decided to convert two of these aircraft to D.560 standard (sports aeroplane). The project was abandoned, and the aircraft were scrapped.

D.551
1/72 scale.

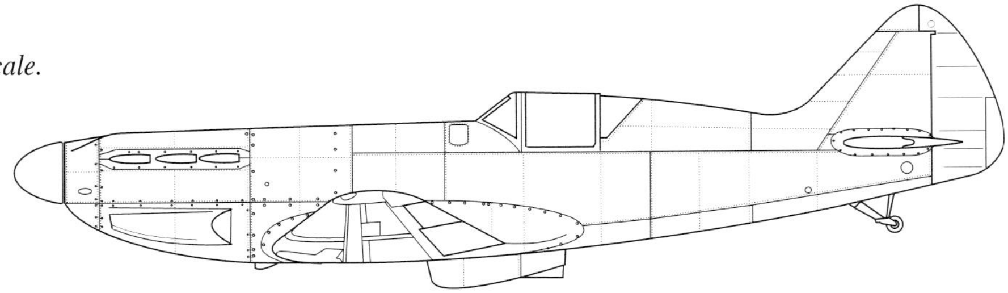

D.553/D.554

Planned version of the D.550 with the Hispano-Suiza HS 12Z engine and Szydlowski-Planiol supercharger.

HD.780

Floatplane fighter development of the D.520, powered by the 1,100 hp Hispano-Suiza HS 12Y 51 engine. The aeroplane was fitted with two floats and inverted-gull wings. Construction of the prototype commenced in November 1939 but was aborted in April 1940 in favour of the D.790.

HD.780
1/72 scale.

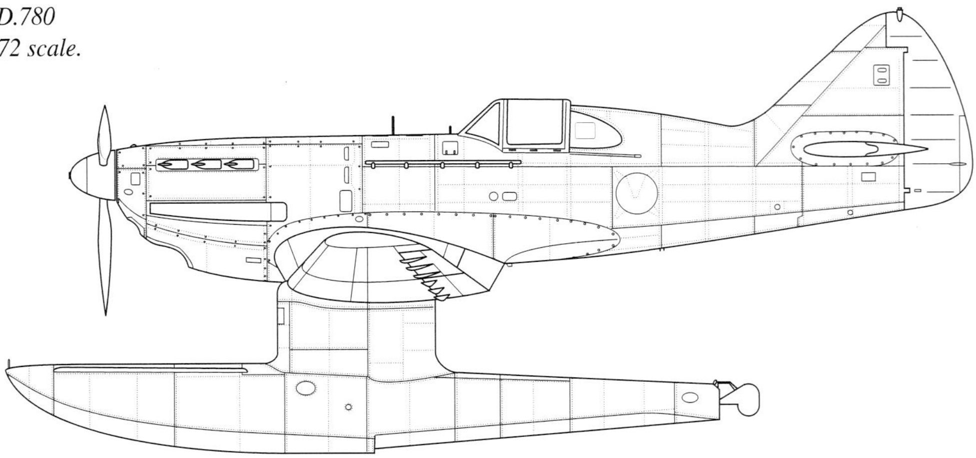

D.790

Planned carrier-borne version of the HD.780 powered by the 920 hp Hispano-Suiza HS 12Y-29 engine, and fitted with D.520 undercarriage. The aeroplane was going to be armed with a 20 mm cannon firing through the propeller shaft, and with two under-wing 7.5 mm guns. Construction was not completed.

D.790
1/72 scale.

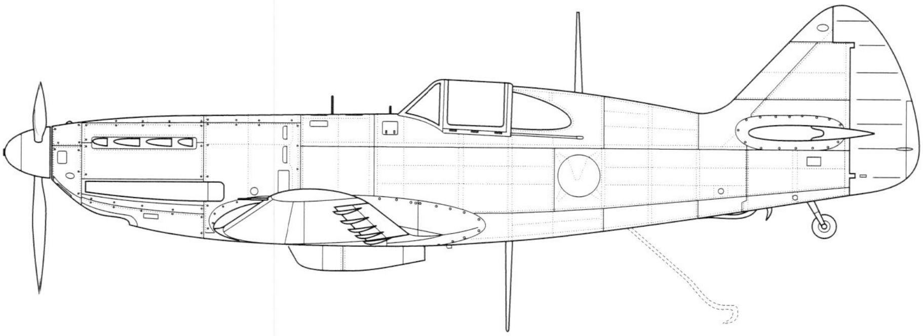

SE.520Z

Development version of the D.520. Modified and lengthened engine cowling, Hispano-Suiza HS 12Z engine. Developed after the liberation, but construction was not started.

Production aircraft

D.520

Production D.520, similar to the third prototype. The aeroplane was powered by the Hispano-Suiza HS 12Y-45/49 engine with the Szydlowski-Planiol supercharger. The first production aircraft was flown on 31 October 1939.

Aircraft nos. 1 to 350 with Hispano-Suiza HS 12Y-45 engines and Ratier 1060 propellers.

Aircraft from no. 351 onwards with Hispano-Suiza HS 12Y-49 engines and Chauviere 3981 propellers.

403 aircraft had been built by the time of the armistice, and 349 more were completed under the Vichy regime.

D.520 DC (double commande)

Post-war conversion of the D.520. The Armée de l'Air ordered a two-seat dual control modification of the D.520. Initially aeroplane no. 243 was converted. Following successful trials, conversion of 20 aircraft was ordered. Eventually only 12 were converted.

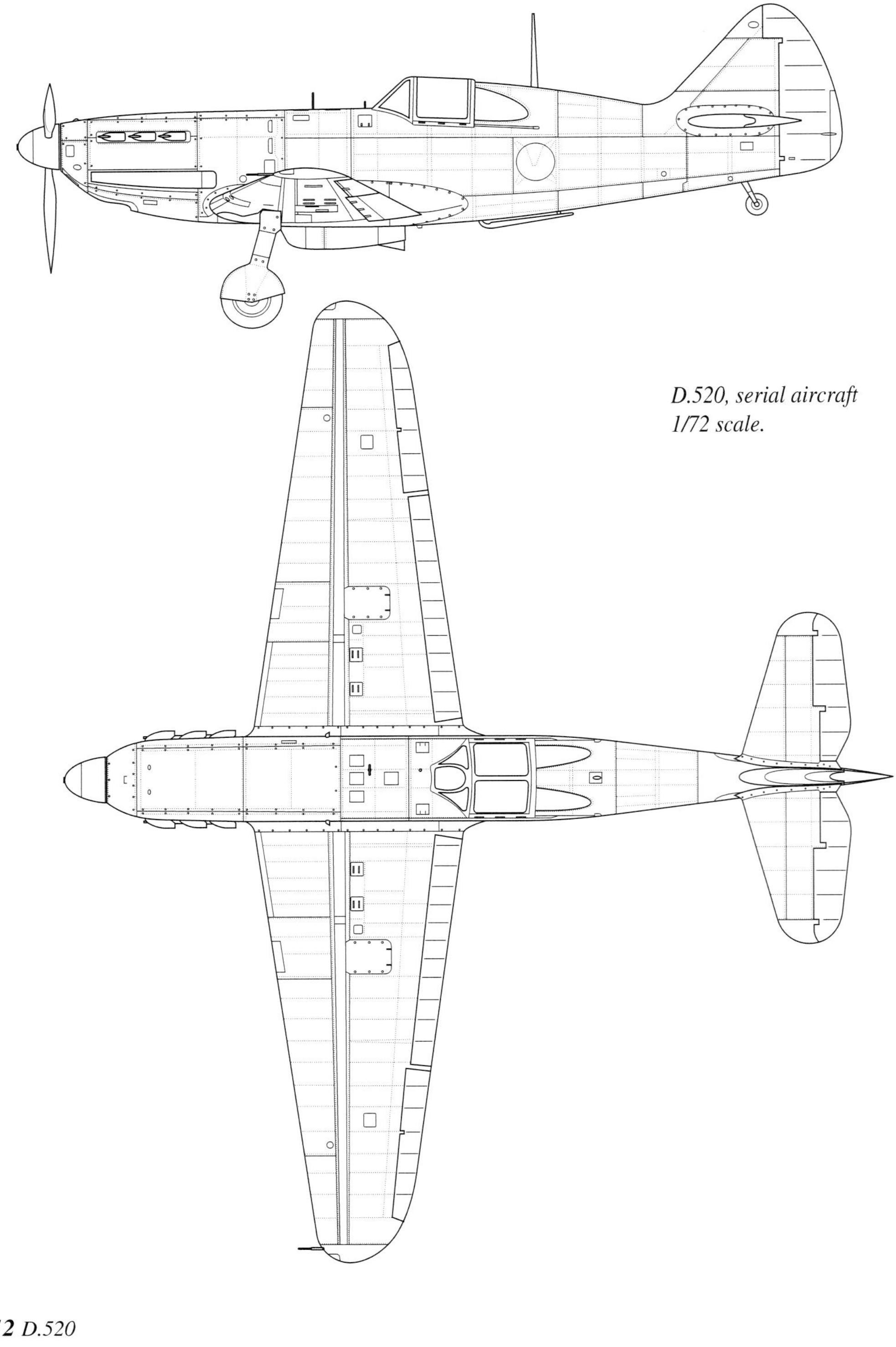

D.520, serial aircraft
1/72 scale.

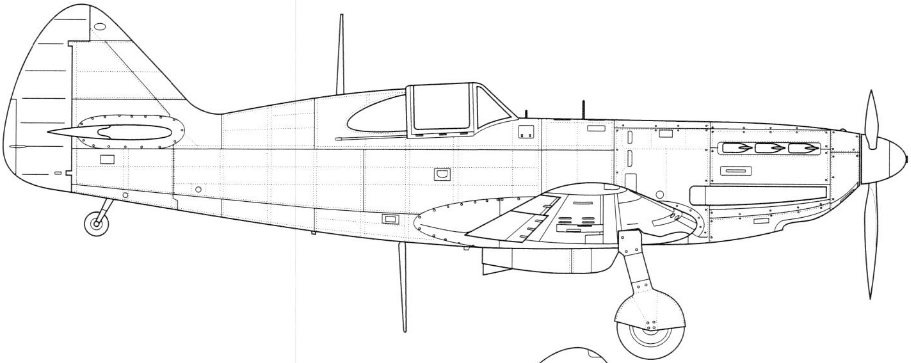

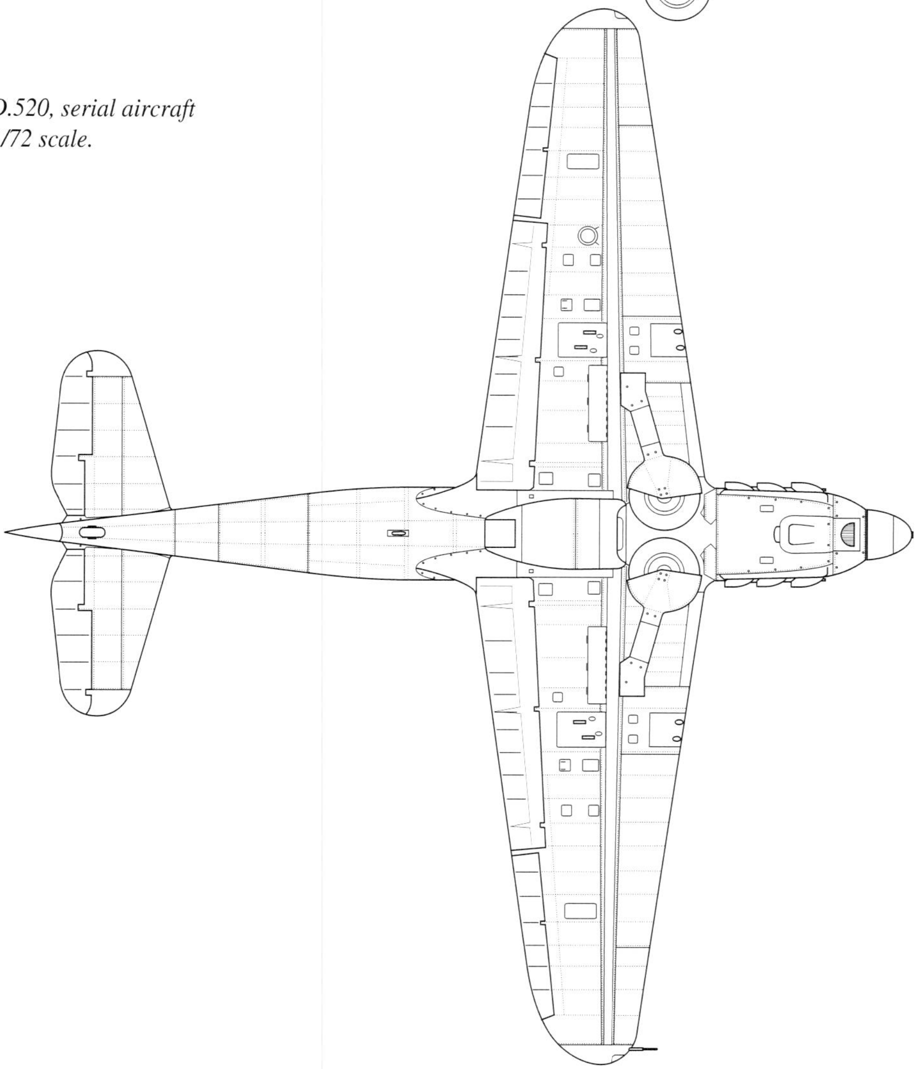

D.520, serial aircraft
1/72 scale.

D.520, serial aircraft, front view.
1/72 scale.

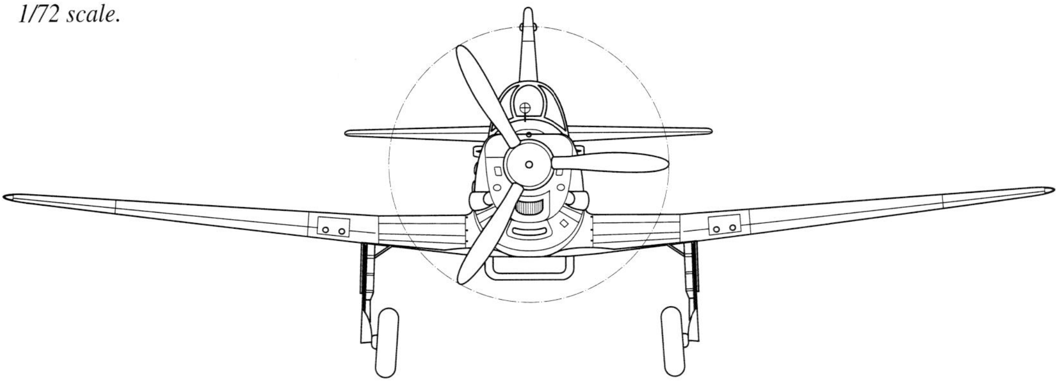

D.520DC
1/72 scale.

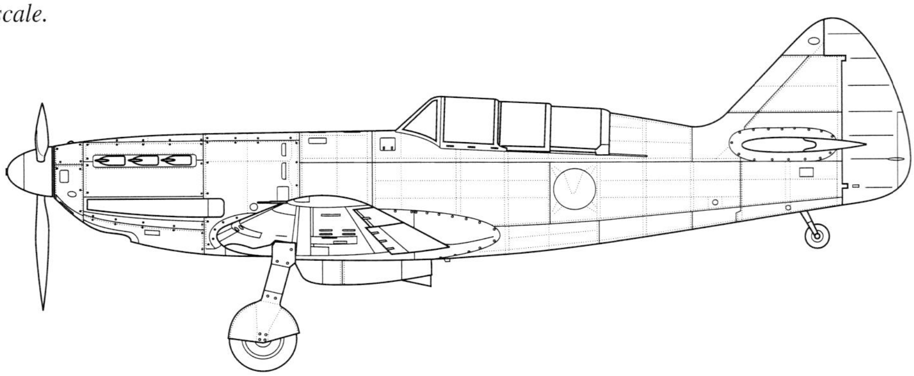

Unidentified destroyed French D.520.
T. Kopański coll.

Western Front 1939-1940

The first production D.520 was not flown until 31 October 1939. At the time of the German invasion of France only a few French units were undergoing conversion onto the new type. On 10 May 1940 few D.520s were in use with front line units. The following squadrons had taken delivery of the type by that time:

- GC III/3, under Cne Bideau, Beauvais-Tillé, 4 D.520s (including 2 operational)
- GC III/6, under Cdt Castanier, Chissey s.-Loue, 3 D.520 (none operational)
- GC I/3, under Cne Thibaudet, Cannes-Mandelieu, 36 D.520 (34 operational)
- GC II/3, under Cdt Morlat, Le Luc, 4 D.520 (2 operational).

GC I/3 was the first to achieve operational capability on 11 May 1940. A few days later GC II/3 received its full establishment of aircraft and these were the first units to completely convert onto the new type.

The D.520 was markedly better than any other French fighter, and was virtually an equal opponent to the Messerschmitt Bf109E. The first encounter with German aircraft took place on 13 May 1940, when GC I/3 D.520s shot down three Henschel Hs 126s and a Heinkel He 111. More French units converted to D.520s during the Battle of France, but lack of time for proper training of the personnel affected the ability to use the D.520. During the fighting of May-June 1940 severe losses were suffered. GC I/3 had written off 24 aircraft, GC II/3 - 16, GC III/3 - 3, GC I/6 – 1, and GC II/7 - 12, including 4 in accidents.

The above listing indicates that also that GC I/6 and GC II/7 re-equipped with the D.520 during the campaign. GC III/6, which received the first

Different shot of the aircraft from page 14.
T. Kopański coll.

D.520 of 3 Esc. GC II/3 shot down over Belgium.
B. Belcarz coll.

machines in early May 1940, fought the campaign on Morane Saulnier MS 406s, only to convert to the new type in early June 1940, when it moved to Le Luc. Ironically, although GC III/6 did not use the new Dewoitine for long, it was a pilot of that unit that became a top ace, using the D.520 to destroy seven out of twelve aircraft that he shot down in the 1940 campaign. On 13 June 1940 while patrolling near Toulon he shot down two Italian Fiat BR.20 bombers, and two days later he became a hero of all France. This was described in his memoirs by a Polish pilot posted to GC III/6, Sgt Michał Cwynar. This is his account:

"Every day two hours of lunch break. Except for a single section of Dewoitines, two pilots at readiness to take-off, all the pilots of the Groupe de Chasse travelled by lorry ca. 4 km to the restaurant at Le Luc. One day the section leader and me had the morning readiness. Pierre asked the flight commander, saying there was no need to leave our section at the airfield during the lunch break because two pilots, W/O Pierre Le Gloan (who had shot down two aircraft near Toulon a few days ago) and another pilot, are

First D.520s were sent to combat units too late to play a major role in the French Campaign.
P. Skulski coll.

D.520 – photo taken from under the fuselage of an MS.406 during the French Campaign.

SHAA

currently patrolling our area. Having obtained the permission of the Escadrille CO, my section leader said we were going for lunch, too.

Pierre Le Gloan was flying in the Toulon area but he had no radio contact with the aerodrome. Leaving the airfield without air defence seemed rather reckless to me, but as a sergeant-pilote and wingman I had no control over this decision. I climbed onto the lorry and we drove to Le Luc for lunch. A French lunch is not only a splendid meal. It is a two hour routine, social relaxation. This time, at the U-shaped table, before the Groupe de Chasse CO at the head of the table had the chance to stand up and raise the toast Vive la Republique!, we heard the slowly approaching, pulsating, monotonous roar of aero engines! We ran into the restaurant's garden. A large group of the Italian Fiat CR.42 biplane fighters formed in echelon starboard was about to strafe our airfield. Incredible scene, shame! But, to a degree, the honour of France was saved. W/O Le Gloan had a hunch and luck, he approached the aerodrome from the direction of Toulon. From a left turn he attacked the last Fiat with such an excessive speed that, in order not to overtake it, he kicked his Dewoitine to the left, to the right, got onto the Italian's tail, and with a short

Fitters of 3 Esc. GC II/7 close to a D.520, the personal aircraft of S/C Jean Doudies.

S. Joanne coll.

burst of his 20 mm cannon and four 7.5 mm machine guns he literally tore the Fiat CR.42 to pieces. The pilot baled out. In the same manner Pierre Le Gloan shot down another Fiat, and a third one. The Italians scattered, started to run! In this sortie the French W/O shot down five enemy aircraft.

From the restaurant we hurried in our lorry to the aerodrome, fearing that our new Dewoitines were strafed. It transpired that only three Morane 406s, no longer used, were slightly damaged. All the Dewoitine 520s were untouched!"

S/Lt Pommier-Layrarouges of GC II/7 was another pilot to achieve significant success in the D.520, when on 5 June 1940 he shot down the Luftwaffe ace Hptm. Werner Mölders. He could not celebrate his victory, though, as he fell to the guns of another Bf 109 and was killed.

The last D.520 episode of the campaign was the evacuation by air from France to the African colonies. The D.520 and the Curtiss H 75A were the only fighters able to fly the route. This episode is not well known, but accounts of these flights include more emotions than those of combats against the Luftwaffe. These ferry hops ended on last drops of fuel, landing on the beach or ditching off the African coast.

***Above*:**
D.520 at factory airfield, Toulouse, 28 June 1940.
P. Skulski coll.

***Left*:**
Duke of Windsor during his visit to GC I/3 at Cannes-Mandelieu.
J. Rajlich coll.

***Bottom*:**
D.520 Serial 129 of GC I/3 on 23 May 1940.
J. Rajlich coll.

***Right**:*
Npor. Jan Eermak, Czech pilot of GC III/3. Photo was taken after the flight to North Africa.
J. Rajlich coll.

***Middle**:*
Aircraft of GC III/3 after arrival in Africa, late June 1940, Relizane airfield.
J. Rajlich coll.

***Bottom**:*
D.520 of 3 Esc. GC II/7, June 1940. Personal aircraft of sgt. Henri Grimaud.
S. Joanne coll.

Three photos of D.520 in Aeronavale markings.
M. Souffan coll.

Above:
Two D.520s of GC II/3 used in Syria against British forces, summer 1941.
B. Belcarz coll.

Right:
D.520 Serial 248 of 2 Esc GC II/7 in Vichy camouflage.
M. Souffan coll.

Bottom:
D.520 nr 368 of GC III/6 at Brindisi airfield, Vichy Air Force.
T. Kopański coll.

Vichy air force 1940-1941

According to the armistice of 22 June 1940 the French air force was going to be reduced, and its deployment was subject to detailed control by the Germans. At the same time in French North Africa chaos reigned, as the military commanders hesitated whether to join de Gaulle or obey their government (and the Germans). The situation was complicated when on 3 July 1940 the British carried out the operation "Catapult" to eliminate the French fleet from possible operations on the German side. British offensive operations resulted in worsened relations between the ex-allies, and in retaliation French Martin 167s flew a raid against Gibraltar on the night of 4 July.

The following Dewoitine D.520 units were deployed in North Africa: GC I/3 (the unit was renamed GC III/3 following a defection of three pilots to Gibraltar on 14 October 1941), GC II/3, GC III/6, GC II/7 and the Aeronavale Escadrille 1AC (ex-AC 1).

Top*: D.520 of 1 Esc. GC I/3, Vichy Air Force.* *M. Souffan coll.*
Middle, left: *D.520 Serial 31, "6" of 1 Esc. GC I/2 at Châteauroux airfield, early 1942. Personal aircraft of Sgt. Lambert. Under the canopy the personal name is visible – 'Chatelia'.* *M. Souffan coll.*
Above*: Aircraft of GC I/2 at Châteauroux. The first two aircraft with Chauviere propellers, the next one with Ratier propeller. M. Souffan coll.*

Above*:*
D.520 of GC I/2, 1941/42.
M. Souffan coll.

Top:
D.520 Serial 277 of GC III/6. Personal aircraft of Lt. Pierre Le Gloan. Photo was taken before the Syrian Campaign.
M. Souffan coll.

Below:
D.520 Serial 229 of GC III/6, Syria, June 1941. Personal aircraft of Cne. Jacobi.
T. Kopański coll.

Syrian campaign 1941

Syria remained in French hands. The increased British activity in the area (suppression of the pro-German rising in Iraq – April 1941 and invasion of Syria – June 1941) resulted in a decision to move another squadron from Maison Blanche at Algiers to reinforce the French forces in Levant (Syria). GC III/6 was ferried by air via Tunis – Catania – Brindisi – Athens – Rayak in late May 1941. It was followed to Syria by GC II/3 (14 June 1941) and 1 AC (25 June 1941). These forces were meant to reinforce the sole French unit in the region, GC I/7 equipped with the obsolete Morane Saulnier MS 406.

The campaign was relatively short and ended in surrender of the Vichy forces in July 1941. By that time GC III/6 pilots scored 24 victories, GC II/3 three, and 1 AC eight more. Between 8 and 14 July all the units withdrew to African bases, with a loss of a total of 40 aircraft.

***Above**:Aircraft of GC II/3, Syria 1941. D.520 "29" belonged to Sgt. Pimont.*
***Below**: D.520 Serial 368, "4" of GC III/6. Personal aircraft of Lt. Brondel. Aircraft was destroyed on 14 June 1941.*

photos T. Kopański coll.

Operation "Torch" 1942

The military operation planned by the Allies called for simultaneous attacks on French naval bases in Morocco, Algeria, and Tunisia. The French forces that defended the territory included a few D.520 squadrons, based at Camp Cazes – Casablanca, Port Lyautey (naval aviation), La Senia, Maison Blanche, and Sidi Ahmed. French reactions varied. Pilots from Maison Blanche did not fight, the others did. The core of the Vichy air force consisted of D.520 squadrons, supported by Curtiss H 75As.

During the short fighting (8-10 November 1942) the Allies lost 44 aircraft, including 25 fighters. French losses amounted to 82 aircraft, including 35 fighters. When on 11 November 1942 the hostilities of Operation "Torch" ended, there were some 160-170 D.520 aircraft in Africa, able to support Allied operations.

At the same time in continental France D.520s equipped GC I/1, GC II/1 and GC III/9, while another 235 were used in training establishment or kept in storage. Most of these fell into the hands of the Germans when they occupied Southern France in late November 1942. Only GC I/2 managed to escape via Corsica to Blida, joining the French forces in Africa.

With the Allies in Africa 1942- 1944

The main problem affecting D.520 operations with the Allies was the different radio system used by the French: HF as opposed to the VHF employed by the Allies. Servicing problems, due to shortage of spares and lack of specialised facilities, meant that in combat conditions D.520 units would suffer high equipment attrition. In late January 1943 it was decided to re-equip French units with Allied types: Spitfire, Hurricane, P-39 Airacobra and P-40 Warhawk. D.520 aircraft were employed for training, a role they fulfilled successfully.

The naval D.520 remained in the front line the longest. A dozen or so aircraft of the type overhauled in late 1942 flew patrols with 1AC (Agadir) and 2AC (La Senia). Both squadrons used a total of 20 D.520s. This state of affairs continued until 1944, when both squadrons re-equipped with other types. They continued to be used for training until 1945, the greatest number being at Meknes. In May 1945 twenty D.520s flew to France, and the following year 19 were shipped by sea. They were sent to the Observers' School at Tours.

F.F.I. – Forces Francais de l'Interieur 1944-1945

The idea to establish air units in liberated France was born in late May 1944 in the district of Hautes Pyrenees in South-Western France. The idea was realised in August 1944, when the area was captured by the Allies. The first unit consisted of two flights without complete establishment, formed at the same time at Tarbes-Ossun and Toulouse. The flights were commanded by the famous test pilot from SNACAM at Toulouse, Marcel Doret. The official name of the squadron was the "Premier Groupe de Chasse FFI de la Region

R4" (The First Fighter Squadron of the Region R4 FFI). The unit was known informally as the "Doret group". The name was used until 1 December 1944, when the unit was reshaped into GC II/18 "Saintonge". At the time it had 15 D.520s. Another reorganisation took place on 1 March 1945, when GC II/18 re-equipped with Spitfire Vs, and its old D.520s went to GCB I/18 "Vendée". This was an unusual unit that included a bomber flight on Dauntlesses and a fighter one on D.520s. Some of the D.520s were allocated to a training unit at Toulouse-Francazal. The aircraft transferred to GCB I/18 continued to be used until the end of the war on the Atlantic coast of France.

France, after the war 1945–1953

D.520s in French service soldiered on until 1953. They were used mostly as training aircraft. The two-seater dual control conversion of the D.520 (D.520 DC – double commande) was an interesting version. Modifications were made at the Atelier Aeronautique in Colombes. Out of the 20 aircraft on order, eventually only 12 were modified. The last flight in the French air force was made on 3 September by D.520 no. 474.

Two D.520s of GC II/18 "Saintogne", 1944/45.
M. Souffan

Above:
D.520 Serial 544 (SV+GB) of JG 103. Photo was taken in September 1944 at Bruksela-Evere airfield. In the foreground, Polish fitters of 308 Squadron RAF.
Belcarz & Pęczkowski coll.

Luftwaffe 1943-1944

Effectiveness of the D.520 was evaluated by Luftwaffe pilots during May and June 1940. After the unoccupied zone of France was invaded by the Germans in November 1942 they captured many aircraft of the type. It was decided that the D.520 would make a perfect conversion type to German fighters in training units of the Luftwaffe. In early 1943 the first aircraft of the type were delivered to JG 101 (Pau), JG 103 (Chateauroux), JG 105 (Villacoublay), and JG 107 (Essey les Nancy).

The Luftwaffe used hundreds of D.520s. More than 100 were destroyed during training.
T. Kopański coll.

The D.520 was a demanding type of aeroplane that caused many problems to the German fledgling pilots. Between March 1943 and August 1944 111 aircraft of the type were lost in accidents in German service. Another 21 were destroyed in Allied raids. In German service the D.520 underwent some upgrades, including the fitting of the German Revi 16b gun sight.

A lot of confusion is caused by the analysis of the painting schemes of these aeroplanes. In many cases this combined the previous French camouflage with German "fish" style schemes. Most German aircraft had their lower surfaces and fins painted yellow. This is confirmed by Ernst Schröder who flew D.520s with JG 101 at Pau (a letter published in Jet&Prop magazine no. 3/2001). Schröder has also summed up the qualities of the D.520:

"The Dewoitine D.520 was a major challenge for us, German pilots. Its flying characteristics were different from the German aircraft types that we knew. For example, it had low control forces. You could fly it "with two fingers", but this required a perfect feel... It flew aerobatic figures perfectly. We flew the "goat" willingly, despite its dangerous character. For the Germans the pneumatic braking system, unknown to us, was a problem. Brakes were actuated by pushing a button on the control stick. Depending on the rudder deflection, they operated on the starboard or port side, when the rudder was set to neutral both wheels braked simultaneously. But woe if the button was pushed too long. Then the aircraft would usually overturn as a result of locked brakes. Another frequent problem with the aeroplane was undercarriage retraction problems, when it jammed while being raised. The resulting insufficient cooling caused a rise of coolant temperature, the fluid spilling on the windscreen, reducing visibility to nil. Only a powerful "shake" to the aeroplane could make the undercarriage retract (10 m above the ground!!!), and the fear disappeared in a second. But your knees continued to shiver!".

D.520 of JG 101 at Pau airfield, France 1944.
T. Kopański coll

Above:
D.520, DN+VG of JG 105, Villacoublay, late 1943.
J.V. Crow coll.

Right:
Uffz Karl Georg Genth during training flight, close to Pau airfield, spring 1944.
T. Genth coll.

Bottom:
D.520, yellow "84" of JG 101, Pau airfield, 1944. Uffz. Gueldenpfennig in the cockpit.
T. Genth coll.

Top:
Two D.520s of JG 101, Pau, 1944.

Middle:
D.520, white "85" of JG 101, Pau airfield, 1944.

Bottom:
Another D.520 of JG 101.
All photos T. Genth coll.

Top:
Fitters of 308 Sqn RAF close to D.520 (SV+GB) at Bruksela-Evere airfield, 1944, kpr. Jan Makowski in the cockpit.
Belcarz, Pęczkowski coll.

Middle:
Uffz. karl Georg Genth preparing for the next flight in his D.520, JG 101, Pau, 1944.
T. Genth coll.

Bottom:
One of the Bulgarian D.520s just after arrival in Bulgaria in 1943. Aircraft still in German markings as used during the ferry flight.
K. Aleksandrov coll.

Regia Aeronautica 1942-1943

The Italians were able to asses the effectiveness of the D.520 in June 1940. But it was due to equipment problems of the Regia Aeronautica in late 1942 that they looked at the D.520 as an aeroplane interesting enough to introduce it into service with their air force. However, the first D.520s in Italian service were aircraft nos. 33 and 309 that remained in Catania, Sicily, and in Rome due to technical problems back in 1941, when Vichy units moved to the Middle East.

Most D.520s entered service in late November 1942. These were both factory fresh and second-hand machines from the French air force. Most of these aircraft were allocated for defence of Italy from Allied air raids. The Italians took over a total of sixty D.520s. Some were allocated to the Regia Aeronautica in 1943 in exchange with the Germans for LeO 451 bombers.

D.520s were allocated to the following units:

2° Gruppo Intercettori, consisting of 152 Squadriglia at Sarazana and 358 Squadriglia at Metato
13° Gruppo CT, 78 and 82 Squadriglias at Metato
22° Gruppo CT, 359 and 371 Squadriglias at Naples-Capodichino
24° Gruppo CT, 361, 362 and 370 Squadriglias at Metato
59° Gruppo BT, 232 and 233 Squadriglias at Metato
60° Gruppo Intercettori, 234 and 235 Squadriglias at Lonate Pozzolo
161° Gruppo Autonomo CT, 164 Squadriglia at Reggio de Calabre
167° Gruppo Intercettori, 303 Squadriglia at Ciampino.

When Italy joined the Allies on 8 September 1943 their air force still had sixteen D.520s in the inventory.

Above:
D.520 of Regia Aeronautica, Istres, France 1943.
T. Kopański coll.

Below:
D.520s were tested by Italian pilots in early 1943 at Istres airfield.
S. Joanne coll.

Bulgarian AF

by Krasimir Alexandrov

In February 1943 the Bulgarian government reached an agreement with the Vichy French government regarding a deal on Bloch MB.152 fighters in exchange for Bulgarian tobacco. The Germans did not approve the contract, but later they offered D.520s that they considered their war prizes, for which Bulgaria would pay them within the Bulgarian-German clearing agreement. The Bulgarians accepted the terms and they sent an engineer and a technical officer to France.

The contract called for delivery of 120 aircraft with spare parts. The Bulgarians received 96 of these (the first arrived in September 1943 at Karlovo airfield). These went to 2/6 and 1/6 Iztrebitelen Orliak (fighter groups). Almost immediately two yatos (squadrons) with 7 aircraft each were relocated near Sofia: 612 Yato to Vrazdebna airfield and 622 Yato to Bozuriste.

The Dewoitines fought their first combat on 10 December 1943 against 60 P-38s and the same number of B-24s flying towards Sofia. The Americans were intercepted by the 3/6 Orliak (equipped with Bf 109s), and 6 D-520s of the 1/6 Iztrebitelen Orliak from Karlovo. That day they suffered their first loss: Capt. Pavel Lazarov Pavlov led the Dewoitines in the head-on attack but he was heavily injured and crashed.

Ten days later 70 Flying Fortresses (Bulgarian pilots referred to all Allied four-engined bombers as "Flying Fortresses") with 60 P-38 attacked again. The French fighters scored their first victory when poruchik (Lt) Nicola Natsev downed a P-38.

With the advent of 1944 Bulgaria received a modest gift of 28 new Bf109Gs and started to replace the French fighters in 2/6 Orliak.

On 10 January 110 P-38s escorted 220 B-17s against the Bulgarian capital city again. They were intercepted by two orliaks: 3/6 and 2/6 (partly with Messerschmitts). The battle also involved 30 German Bf109Gs armed with four rockets each. According to the memories of Col. (retd.) Petar Kirov Petrov:

D.520, "14" of 6 Orliak at Karlovo airfield, September 1943.

K. Aleksandrov coll.

"Our 2/6 Orliak on 10 January took off in full establishment of 32 Dewoitines. Arranged in a column of fours we climbed to the West. We were at about 7,000 m when we saw the 'Fortresses'. As we were on the approaching course, we went in the frontal attack, like on 20 December. None of us trembled, and none moved away from the steel avalanche approaching us with huge speed. Everyone took one 'Fortress' in the gun sight and waited for the moment to push the trigger on the machine guns and cannon. None of us paid attention how we retreat after the attack. Wing by wing we came near the 'Fortresses'. We saw how they moved up-down, left-right. They were upset by our attempt to scatter them with frontal attack. All in this moment was for seconds: take aim at the last moment, open fire and move away back from the attack in any possible way for everyone. One was running into the battle, another exited by semi looping, third aside and up. The effect of our frontal attack was great: the first group of bombers deviated from the attack, dropped their bombs on a field, without taking the battle course to the heart of Sofia".

In this battle Lt. N. Natsev made a forced landing after a dogfight with six "Lightings", podoficer (Sgt) A. Krastev abandoned his aircraft, but he was shot in the air by American aeroplanes.

Big events (and troubles) were ahead. 30 March 1944 was a clear, cloudless day. 450 B-17s, B-24s, B-25s and Halifaxes with 150 P-38s took course towards Sofia. 73 fighters took off against this "undefeated armada": 20 Bf 109s (3/6), 6 D.520s (1/6), and 28 Dewoitines from Karlovo. Ratio of 8 to 1. Iztrebitelen Orliak 1/6 was late because its airfield had a lot of damage (after the last bomber attack) but nevertheless it took part in the battle. Lt. Bogdan Bogdanov from 2/6 pursued enemy bombers almost 100 km and shot

Hangar at Karlovo airfield, September 1943.
K. Aleksandrov coll.

D.520 of 1/6 Iztrebitelen Orliak.

down one plane. In this fight 2/6 had big loses: 4 aircraft and 3 pilots. On 5 April 600 bombers and 200 fighters flew to Bucharest. 3/6 and 2/6 were scrambled and shot down one aircraft. 12 days later 100 P-47s and P-51s escorted 350 "Flying Fortresses" toward Sofia. 3/6 and 2/6 took off against them, led by Capt. Bosniakov with 16 Messerschmitts and two D.520s, later joined by five Dewoitines from Karlovo. Capt. Bosniakov shot down one bomber, and Lt. Kovatsev another, but that was a "black Friday" for Bulgarian fighters: they mistook Mustangs for Messerschmitts and this allowed the Mustangs to shoot at the Bulgarian fighters from very close distance. Three pilots were killed, one heavily injured and made forced landings. This was the last fight of the Dewoitines.

For Bulgarian pilots the Dewoitine was a "...very manoeuvrable aircraft with excellent cannon with precision and penetrating power of shell."

After the war the number of Dewoitines diminished rapidly and they were soon scrapped.

D.520, "14" of 6 Orliak at Karlovo airfield.
Photos K. Aleksandrov coll.

Technical description

Fuselage

Oval cross-section. Made up of 12 frames (including 4 double ones) connected by four longerons. The engine mount made of welded steel tubes, and the wing spar, were both attached to the first main frame. The main fuel tank was located between frames nos. 2 and 4, separated from the cockpit by the fireproof bulkhead. The radio and luggage compartment was located aft of the cockpit. The last (double) frame supported the tail wheel. The last fuselage frame provided support for the fin.

The fuselage was covered with flat duralumin panels riveted onto the structure. The coolant radiator was located under the fuselage. The round luggage compartment cover was located on the port side of the fuselage.

Cockpit

The hood was single-piece, sliding back. Long windows were added to improve rearward visibility. The manually adjusted pilot's seat with place for the seat-type parachute was protected from behind by a 5 mm armour plate.

Wing

All metal design, single spar, with round tips. The wing spar was attached to the main fuselage frame with four attachments. Each wing consisted of two sections, joined along the spar. Ailerons and flaps of all metal design, duralumin covered. Ailerons with weight balance, pushrod actuated. Flaps hydraulic actuated. Navigation lights at wing tips.

Tail

Cantilever surfaces of all metal design, duralumin covered. Tailplane adjustable in flight. Rudder with horn balance and a trim tab. Elevators axially balanced, pushrod-actuated. Rudder cable-actuated.

Undercarriage

Tail wheel layout, with OLAER single main wheel legs pivoted on the wing spar. Undercarriage retracted inwards into the wing, partly covered with single-piece covers. 650 x 236 mm Goodrich type 10 bis tyres with pressure of 0.36 MPa. OLAER hydraulic shock-absorbers with 300 mm travel. OLAER pneumatic drum brakes. Brake actuator button on the control stick. Castoring fixed tail wheel, with OLAER hydraulic shock-absorber with 67 mm travel and a 220 x 105 mm wheel fitted in a fork.

Power plant

Hispano-Suiza type 12Y45 36-litre 12-cylinder Vee in line engine, rated at 688 kW, 2,520 rpm for take-off and at 648 kW, 2,400 rpm for normal operation. Liquid cooled (mixture of ethylene glycol and water). Dry weight of the engine alone 515 kg.

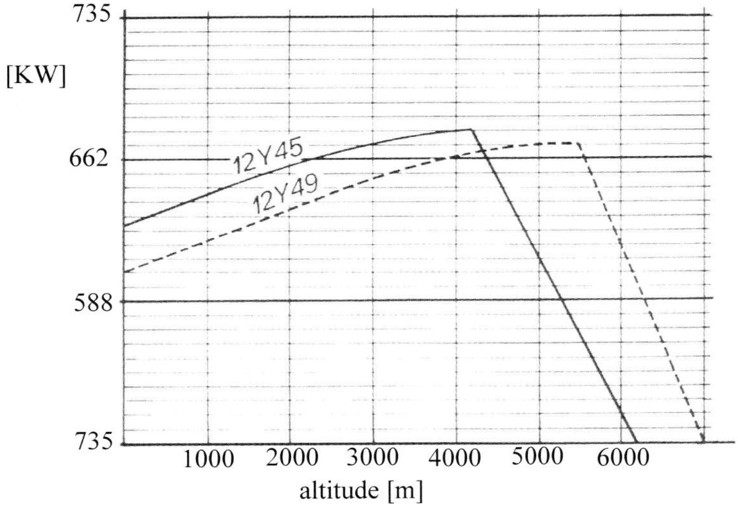

Engine fittings

Two RB type P12A magnetos, six Solex type 56SVC carburettors, Szydlowski-Planiol supercharger and Hispano altitude corrector. The engine drove a 3,000 mm Ratier type 1606M or Chauviere typ 3981 three-blade all metal constant speed propeller. The Ratier propeller featured an electric pitch control, while the Chauviere was pneumatic. Closed circuit engine cooling by a radiator under the fuselage. VIET type 250 pneumatic starter with a VIET compressor.

Fuel system

92 or 100 octane aircraft petrol. 396 litre main tank in the fuselage and two 120 litre tanks in outer wing panels forward of the spar. Total fuel capacity in the aeroplane: 636 litres. Fuel circulation forced by two AM type fuel pumps driven by the engine.

Oil system

Oil pump, 58 litre oil tank (filled to 45 litres) and oil cooler located in the air duct in lower fuselage.

Electric system

Single wire 24 V. Current supplied by the 1,200 kW LABINAL generator and 20 Ah SAM battery.

Hydraulic system

Used to lower the main undercarriage and to actuate the screens in the coolant radiator and oil cooler, and to fold the ventral aerial. Fed by the OLAER hydraulic pump driven by the engine. Pressure in the system 25.5-27.0 MPa.

Pneumatic system

VIET compressor driven by the engine, an air cylinder for the starter, and several other cylinders used to actuate the triggers and reloading of the weapons, the pitch control mechanism, and the main wheel brakes.

Oxygen system

1,000 litre oxygen cylinder located aft of the armoured plate behind the pilot's seat, used for cockpit ventilation and to feed the MANURELLE or GORDOU oxygen mask.

Radio system

RADIO INDUSTRIE type 537 HF transmitter-receiver, transmitter aerial (fixed) on the top of the fuselage and receiver aerial (folded) under the fuselage. Radio located aft of the pilot's seat.

Additionally, the aeroplane was equipped with a first aid kit, automatic fire extinguisher on the port side of the engine, and an armament anti-freeze system in the wings.

Armament:

20 mm HISPANO SUIZA type HS 404 aircraft cannon located between the engine cylinders and firing through the propeller shaft, fed from ammunition belt in a drum on the weapon (60 rounds).

Four 7.9 mm MAC 34 type M39 machine guns (two in each wing) with 675 rounds in belts.

OPL RX reflector gun sight, ring-and-bead sight on the fuselage.

According to a CEMA report of 3 October 1939, official tests of the production aeroplane no. 02 carried out between 29 August and 10 October 1939 gave the following results for the aeroplane with the HSY45 engine, Ratier 1606 propeller, weighing 2,567 kg and fed with 85 octane fuel:

- maximum speed at 5,400 m: 535 km/h
- maximum ceiling: 10,250 m
- maximum range (main tank only): 890 km
- maximum range (main and wing tanks), at cruising speed of 370 km/h at 5,100 m: 1,540 km
- take-off run: 370 m
- landing run: 480 m

Production aircraft performance:

Max. speed	at an altitude of	Time to height
(km/h)	(m)	
495	1,000	1'22"
495	2,000	3'17"
520	3,000	4'02"
545	4,000	5'13"
550	5,000	6'41"
542	5,200	8'16"
473	6,000	13'24"
	8,000	16'58"

engine revolutions (rpm)	speed (km/h)	fuel consumption (l/h)
2400	540	315
1775	355	110

Selected bibliography:

Bączkowski W. – Dewoitine D.520, Books International, Warsaw 1998

Belcarz B. – Polskie lotnictwo we Francji 1940, Stratus, Sandomierz 2002

Cwynar M. – Wartime Recollections, Warsaw 2004

Danel R., Cuny J. – Le Dewoitine D.520, Docavia, Paris 1972

Danel R. Cuny J. – Les Avions Dewoitine, Paris 1980

Duning C., – Courage Alone, Hikoki, Aldershot 1998

Marechand P., Takamori J. – Le Dewoitine D.520, Along, Le Muy 2002

Martin J., Martin P. – Ils étaient lŕ… L'armée de l'Air septembre 39 – juin 40, Aero Editons, Fleurance 2001

Listemann P.,Tilley P.A., Ehrengardt Ch.J. – Les pilotes de chasse français 39-45, Aero Editions, Fleurance 1999

Rajlich J. – Na nebe sladke Francie, Ares, Praha 1997

various authors – Dewoitine D.520, DTU, Bruxelles 1997

various authors – Fighters over France and Low countries, MMP/Stratus, Sandomierz 2002

Avions, Air Magazine, Aero Journal, Jet&Prop

GENERAL VIEW

Three photos of preserved D.520 in flight.

J. Kightly

***Above*:**
D.520 during landing.

***Right*:**
D.520 in flight.

***Bottom*:**
D.520 in inverted flight. Note the undersurfaces of the wings, rather a rare view.

All photos J. Kightly

FUSELAGE

Above: *Very nice shot of the D.520 port fuselage from the rear.*
Below: *Starboard side of the fuselage, almost plan view.*

P. Got.

Above: *Front, port fuselage of D.520*
Below: *Starboard, rear part of the fuselage. P. Got*

Left:
Two photos of the D.520 in the Le Bourget Museum.
B. Belcarz

Bottom.
Starboard side of the D.520 fuselage.
P. Got

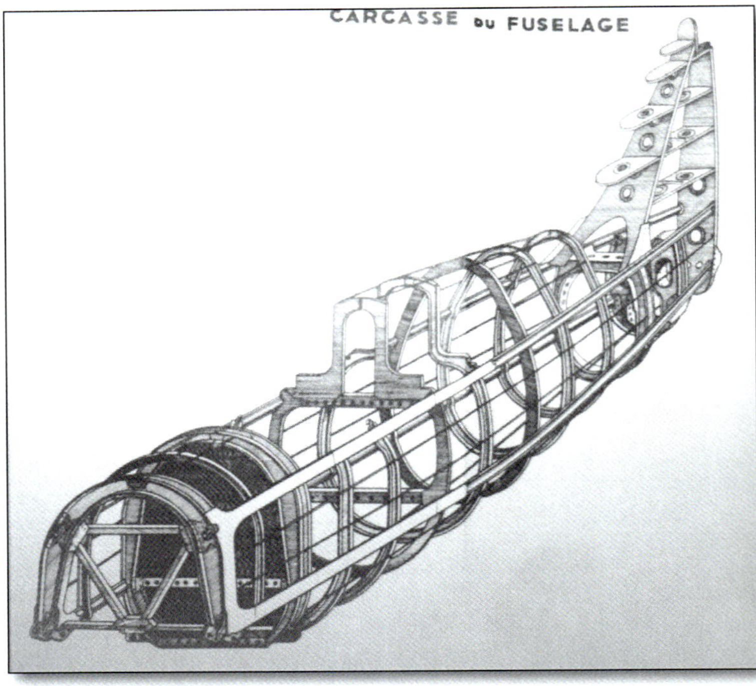

Above:
D.520 during restoration, still unpainted.
P. Got.

Right:
D.520 fuselage construction. Drawing from the Technical Manual.

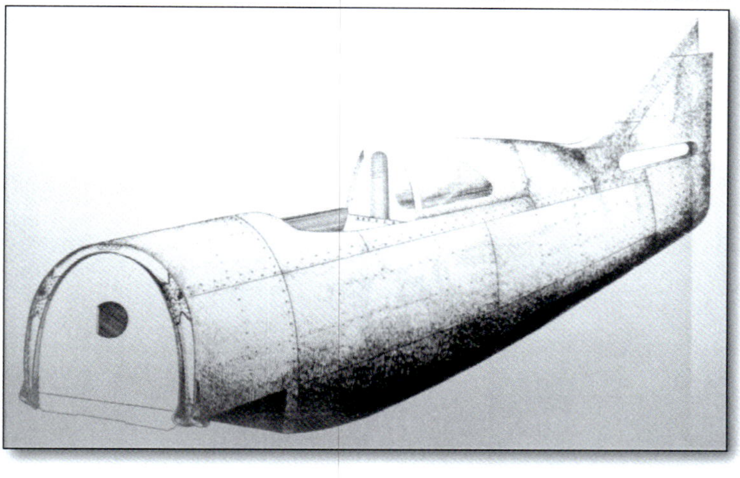

Top and middle photos:
Another two photos of D.520 during restoration.
P. Got

***Bottom**:*
Fuselage skins. Drawing from the Technical Manual.

Above and right:
D.520 fuselage during restoration.

P. Got

***Bottom**:*
Middle, port side of the fuselage.

J. Kightly

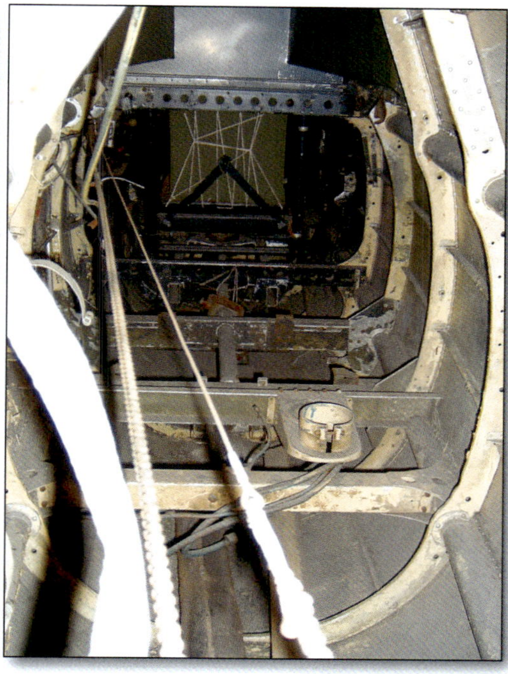

Above, left*: Middle, front part of the fuselage.* ***Above, right****: Fuselage frames. J. Kightly*
Below*: Starboard, rear part of the fuselage and the fin. P. Got*

WING

Top: *Starboard wing from the rear.*

J. Kightly

Bottom: *Wing planform from above, rear.*

P. Got

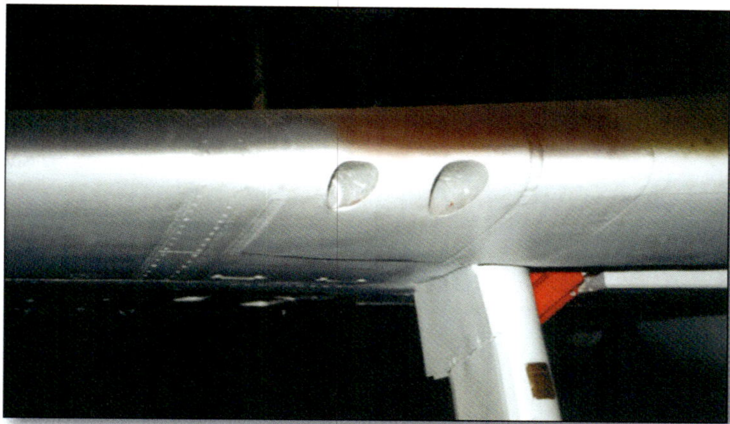

Top:
Machine gun ports in the starboard wing.
B. Belcarz

Middle:
Underside of the starboard wing.
J. Kightly

Bottom:
Shell ejection chutes under the starboard wing.
J. Kightly

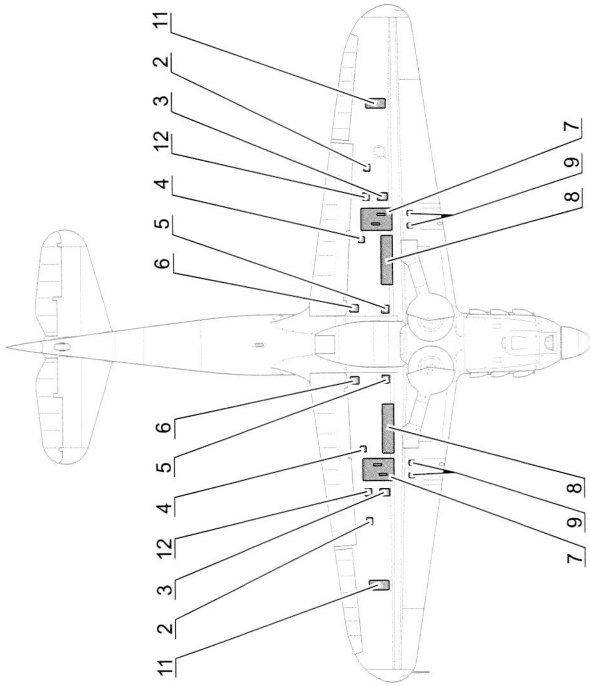

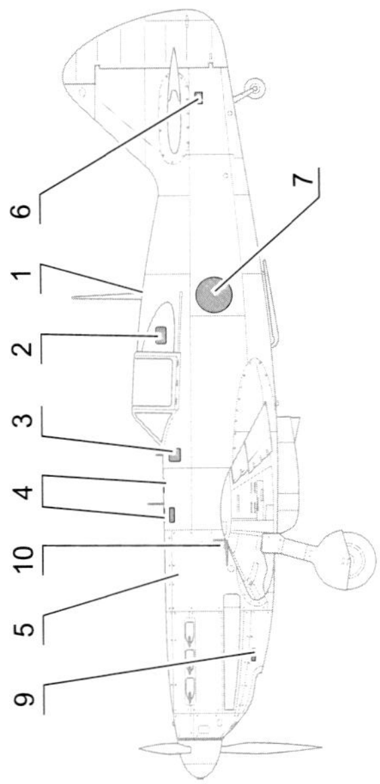

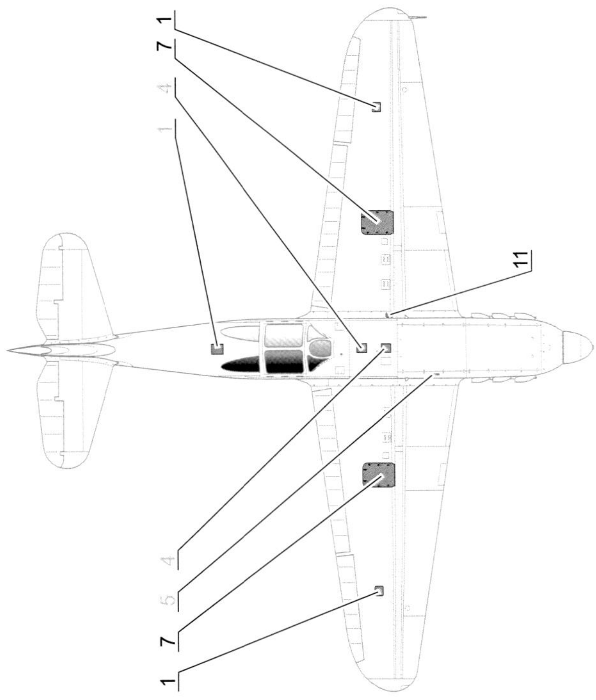

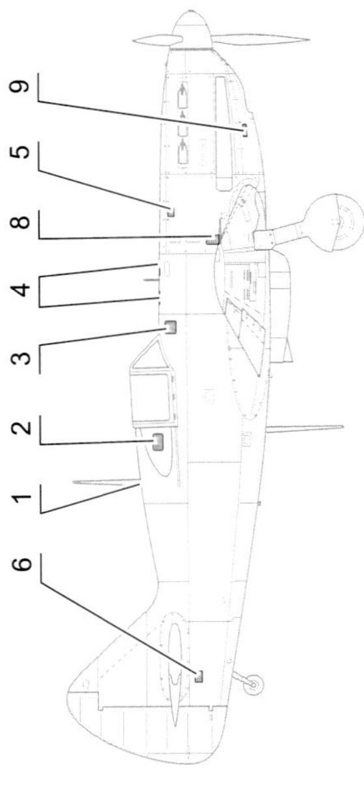

***Opposite page**:*
Inspection panels

	Wing		***Fuselage***
1.	*Aileron control rod access panel*	1.	*Access to aerial connector*
2.	*Aileron control rod access panel*	2.	*Inspection panel*
3.	*Access panel*	3.	*Access to pilot instrument panel*
4.	*Aileron control rod access panel*	4.	*Fuel filler caps*
5.	*Aileron control rod access panel*	5.	*Oil filler cap*
6.	*Landing flap control rod access panel*	6.	*Access to tail wheel shock absorber*
7.	*Machine guns access panel*	7.	*Luggage compartment*
8.	*Ammunition loading*	8.	*Access to oil cooler*
9.	*Auxilliary access panel*	9.	*Access to oil dump cock*
11.	*Aileron control rod access panel*	10.	*Access to air cock*
12.	*machine guns heating door*	11.	*Inspection panel*

Starboard side of the fuselage.

P. Got

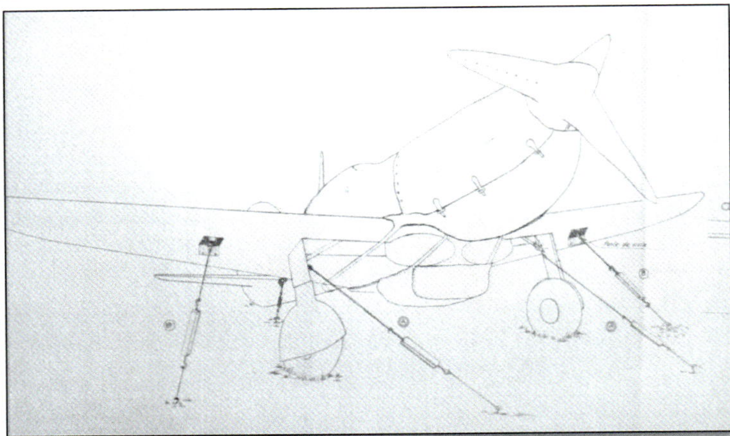

Above:
Port wing machine guns.
J. Kightly

Above right:
Drawing from the Technical Manual showing how to secure aircraft on the ground.

Right:
Port wing machine gun shell ejection chutes
J. Kightly

Bottom:
Front view of the wing-fuselage joint section
J. Kightly

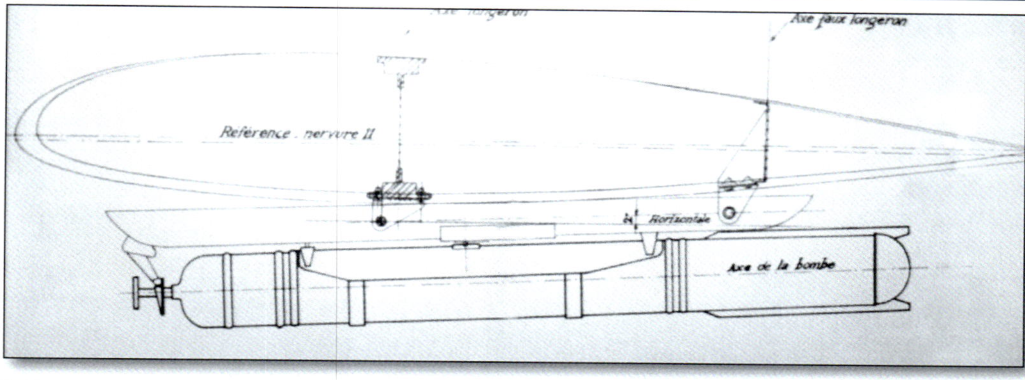

Top:
Bomb under the wing. Drawing from the Technical Manual

Middle:
Port wing tip showing the base of the pitot tube and the red navigation lamp.
J. Kightly

Bottom:
Port aileron, underside view.
J. Kightly

Detail photos

Drawings from the Technical Manual showing the wing structure.

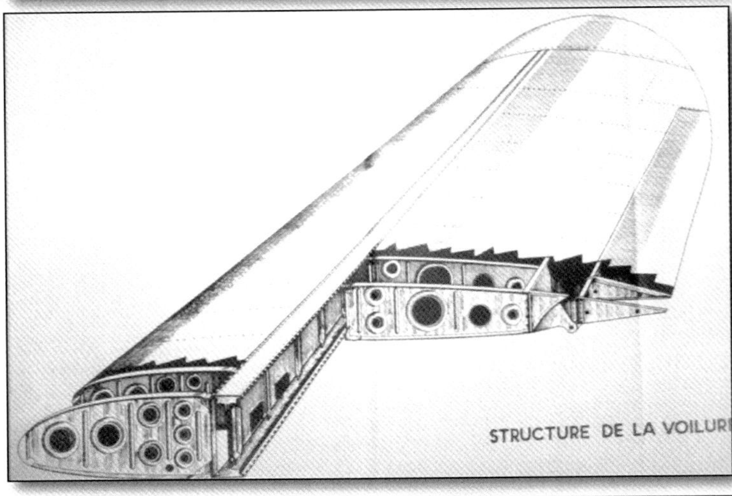

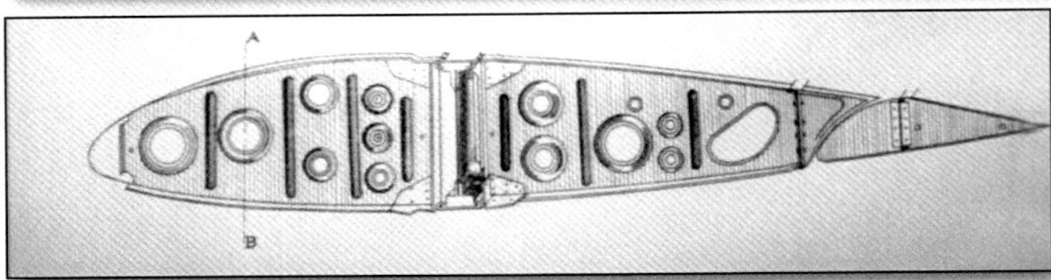

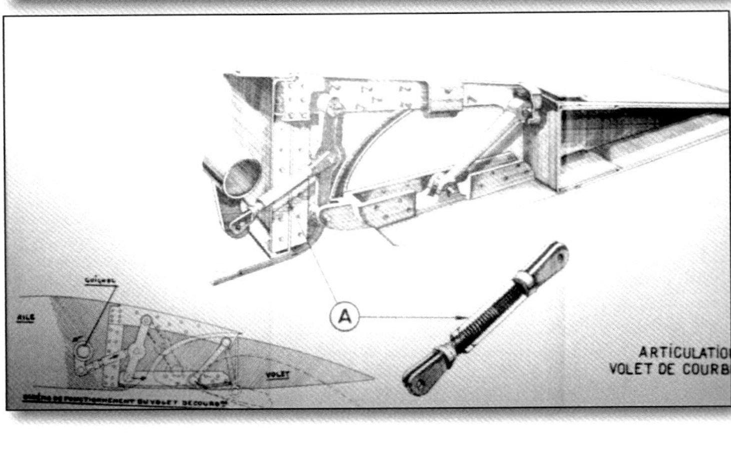

Top:
D.520 in storage
J. Kightly

Middle:
Undersurfaces of the port wing.
J. Kightly

Bottom:
D.520 during restoration. Note the landing flap structure.
J. Kightly

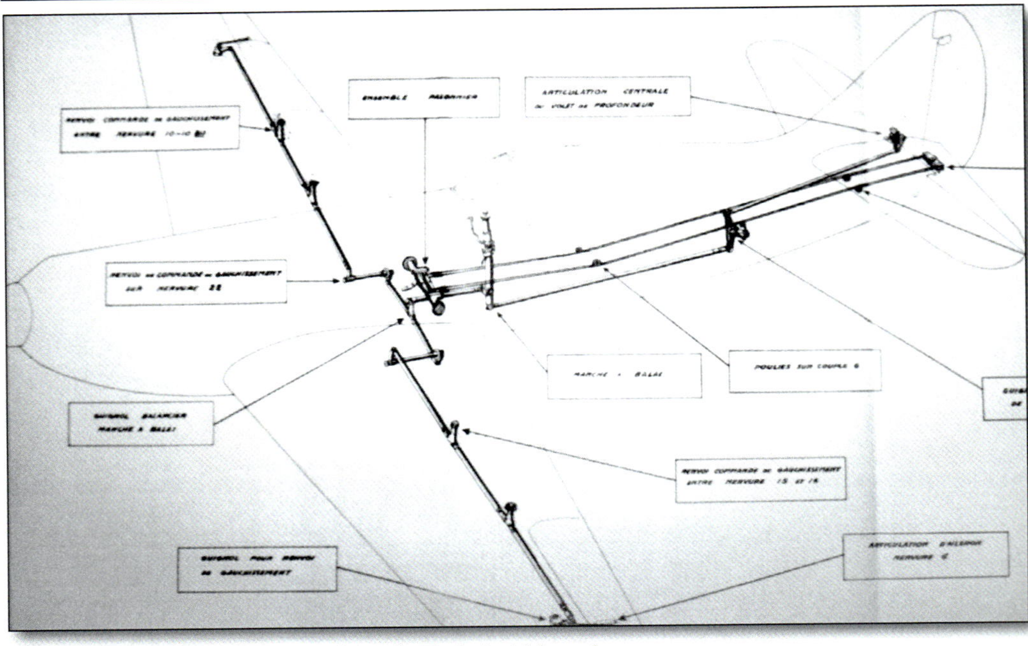

***Above**: Control system. Drawing from the Technical Manual*
***Below, right:** Undersurface of the wing centre section. J. Kightly*
***Below, left:** Fuselage-wing fillet, front view. J. Kightly*
***Bottom, right**: Starboard wing. J. Kightly*

CANOPY & COCKPIT

__Above__: Canopy in the evening light, from the rear. P. Got

***Left and below**. Drawings of the canopy details, from Technical Manual.*

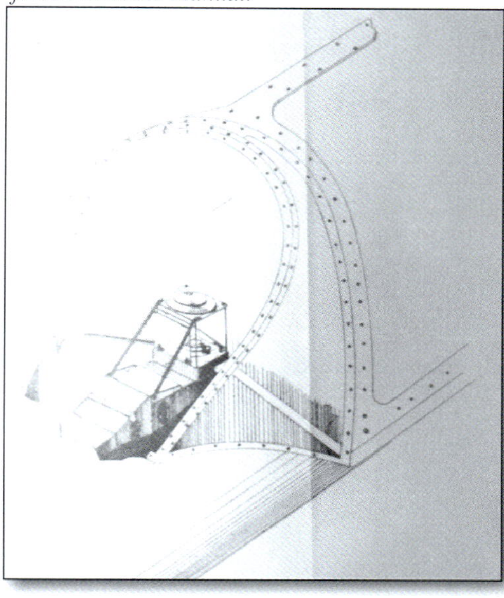

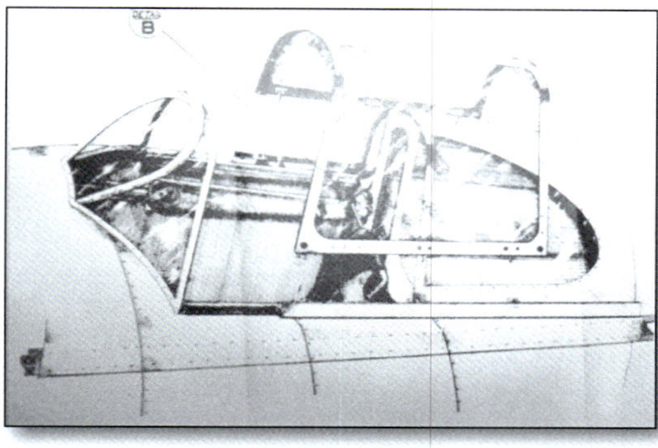

Above:
Detail photo of the canopy from the right
P. Got

Left and below:
Three photos of the canopy during restoration.
J. Kightly

Above: Instrument panel, original wartime photo.

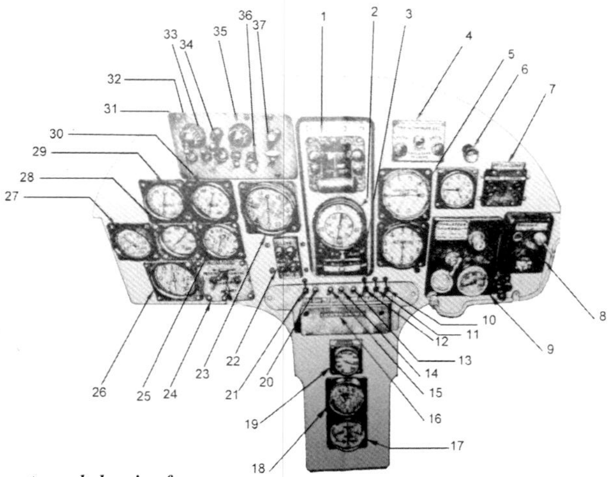

Instrument panel, drawing from the Technical Manual.

1. Compass VION
2. Flight director (controller)
3. Anemometer
4. Bomb release gear
5. Drift indicator
6. Switch
7. Main supply connector
8. Oxygen system relays
9. Oxygen inhaler
10. Port machine gun switch
11. Starboard machine gun switch
12. Arming port machine gun switch
13. Arming starboard machine gun switch
14. Cannon switch
15. Ammunition counters
16. Fuse Box
17. Undercarriage system pressure gauge
18. Brake pressure gauge
19. Fuel gauge
20. Pilot light
21. Camera gun main switch
22. Landing gear position indicator
23. Landing speed indicator
24. Blade angle setting
25. Oil pressure gauge
26. Tachometer
27. Fuel system pressure gauge
28. Carburettor pressure gauge
29. Cooling liquid temperature gauge
30. Oil temperature gauge
31. Navigation lights switch
32. Wing light switch
33-37 Radio and signal switches

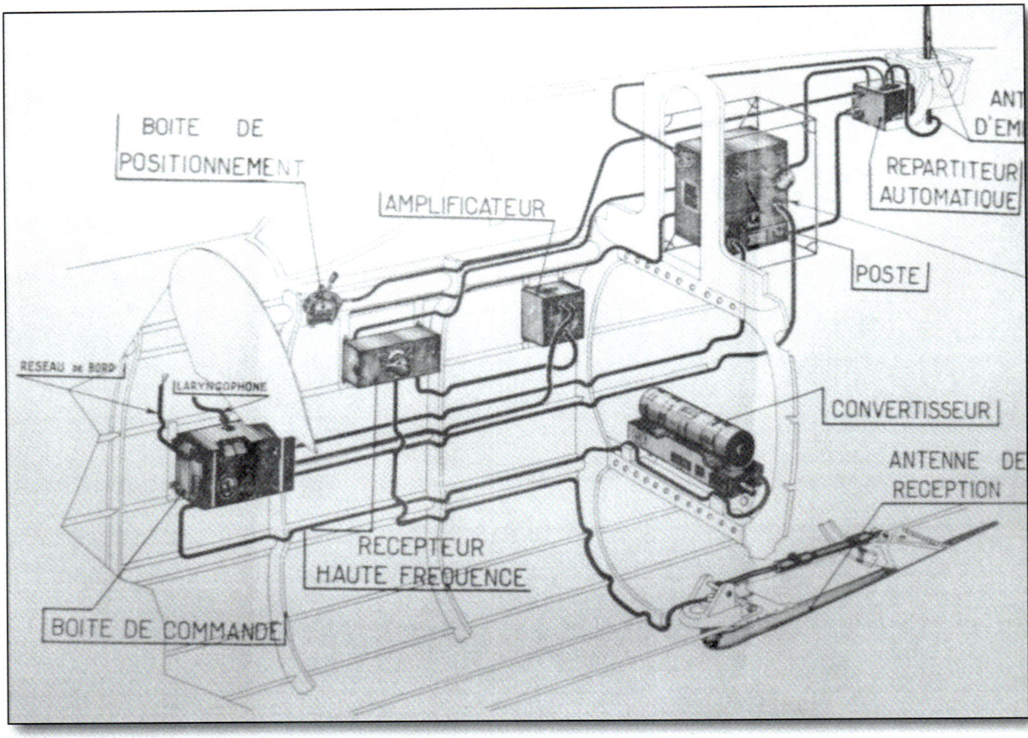

***Above**:*
Radio system, drawing from the Technical Manual.
***Right**:*
Pilot's seat, from the Technical Manual
***Below**:*
Instrument panel during restoration.
J. Kightly

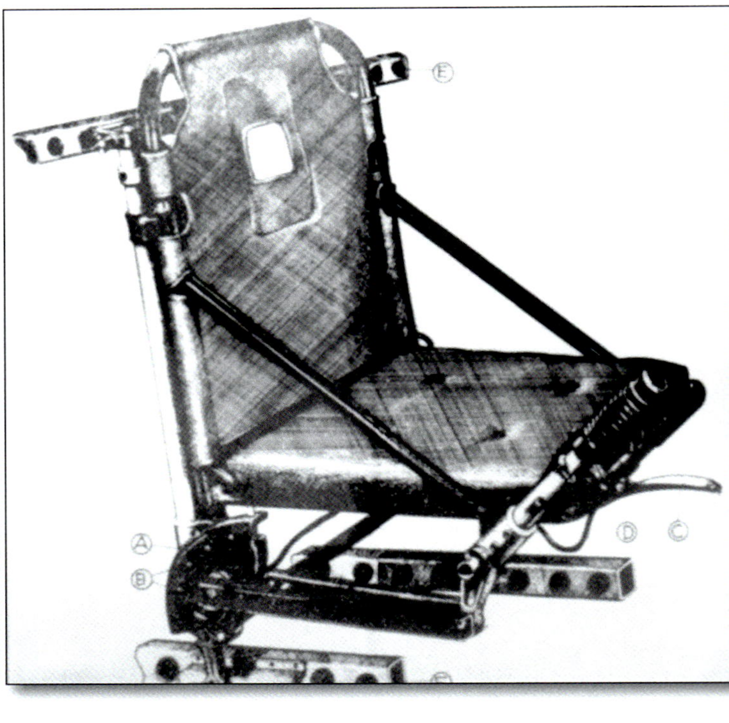

Above and left:
Three photos of the cockpit details.
J. Kightly

Bottom:
Canopy from the left.
J. Kightly

TAIL

***Above**:*
Starboard side of the tail, rear view.
P. Got

***Right**:*
The last fuselage frame.
J. Kightly

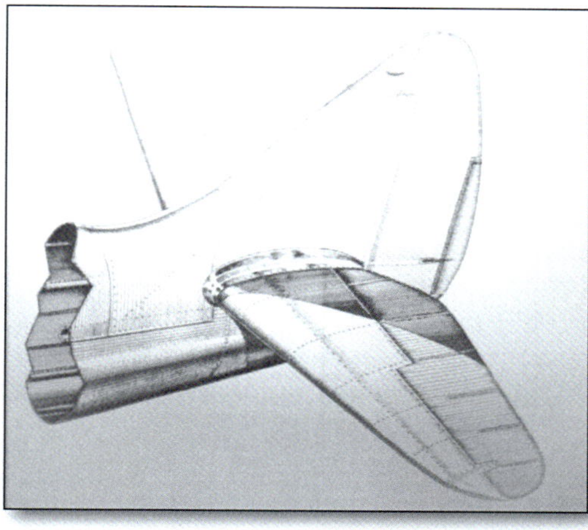

Above, left: *Starboard side of the rudder. B. Belcarz*
Above*: Tail structure, drawing from the Technical Manual*
Left*: Fin during restoration. J. Kightly*
Below*: Rudder structure, from the Technical Manual.*

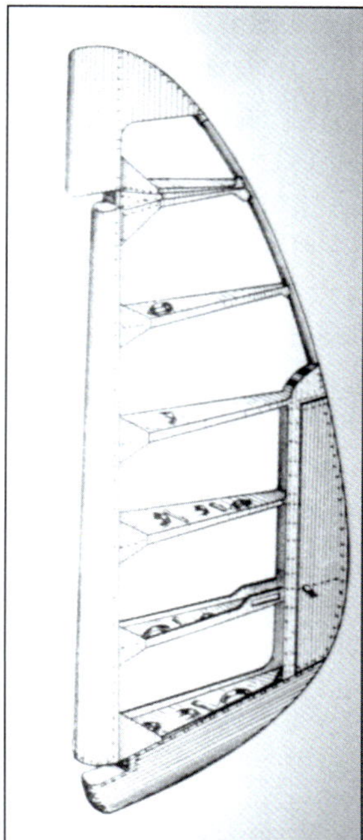

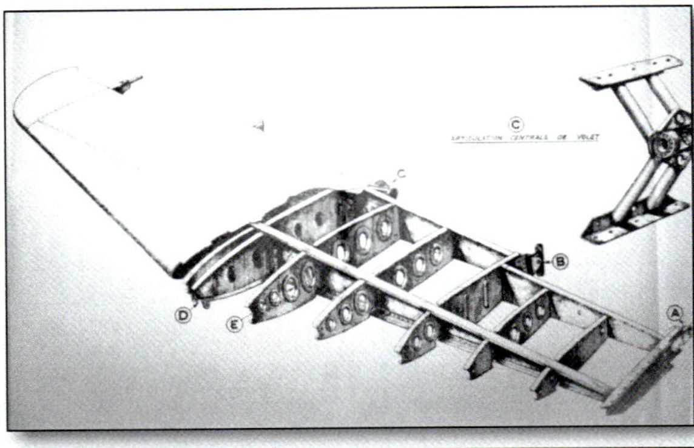

Opposite page:
Two photos of the engine details and engine mount.

P. Got

Top, left. *Horizontal stabilizer structure. Drawing from the Technical Manual.*
Top, right: *Navigation light on the fin tip. J. Kightly*
Middle: *Tail during restoration. Rudder still uncovered. J. Kightly*
Above: *Starboard side of the tail. T. Genth*
Right: *Rudder details. B. Belcarz*

ENGINE

Above:
Engine almost from the front. Note that engine lacking many details and piping.

P. Got

Right:
Engine mounting. Drawing from the Technical Manual

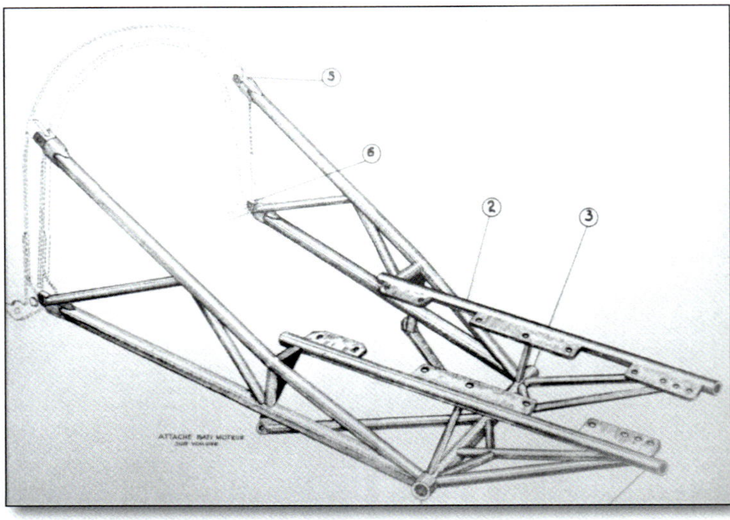

***Right**:*
Details of the engine covers, starboard side.
B. Belcarz

***Below**:*
Engine from the right, front view.
J. Kightly

Bottom:
Engine with all covers removed.
P. Got

Three photos of the spinner details. Note also details of the front engine covers.
Photos B. Belcarz (1) and J. Kightly (2)

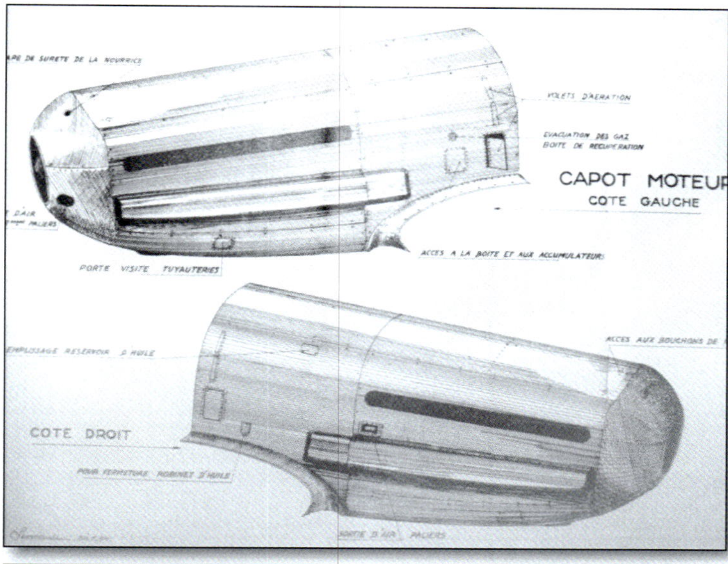

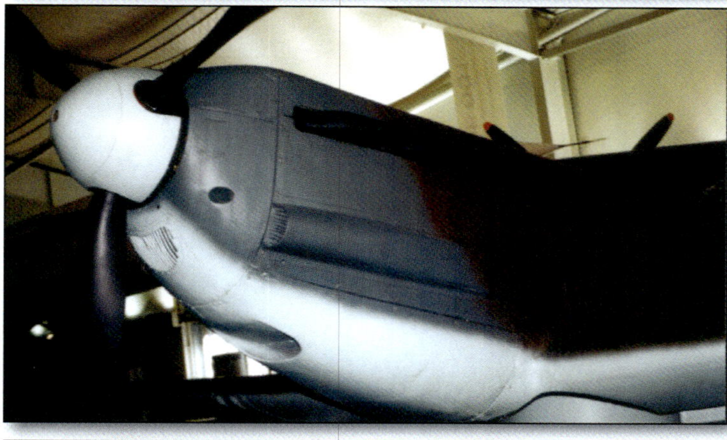

***Top**:*
Drawing from the Technical Manual showing the engine covers.

***Middle**:*
Portside, front part of the engine covers.
B. Belcarz

***Bottom**:*
Front part of the fuselage.
M. Lowe

Details of the D.520 engine.
Oil cooler.
Exhaust pipes.
Photos J. Kightly

Above: *Upper part of the engine.*
Right: *Oil tank.*
Below: *Two photos of the engine – front view.*

Photos J. Kightly

Detail photos

Top:
Engine cooler.
J. Kightly

Middle:
D.520 – front view. Note the details of the propeller.

Bottom:
Propeller boss.
J. Kightly

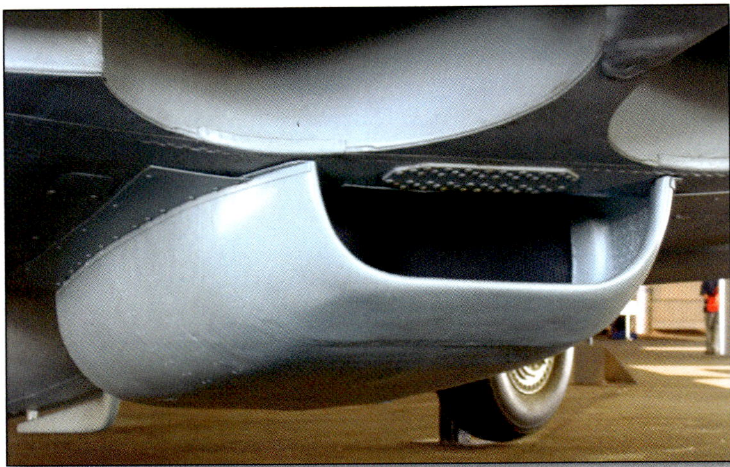

Above: *Two photos of the oil cooler inlet. J. Kightly*
Below: *Drawing from the Technical Manual showing engine starting system.*

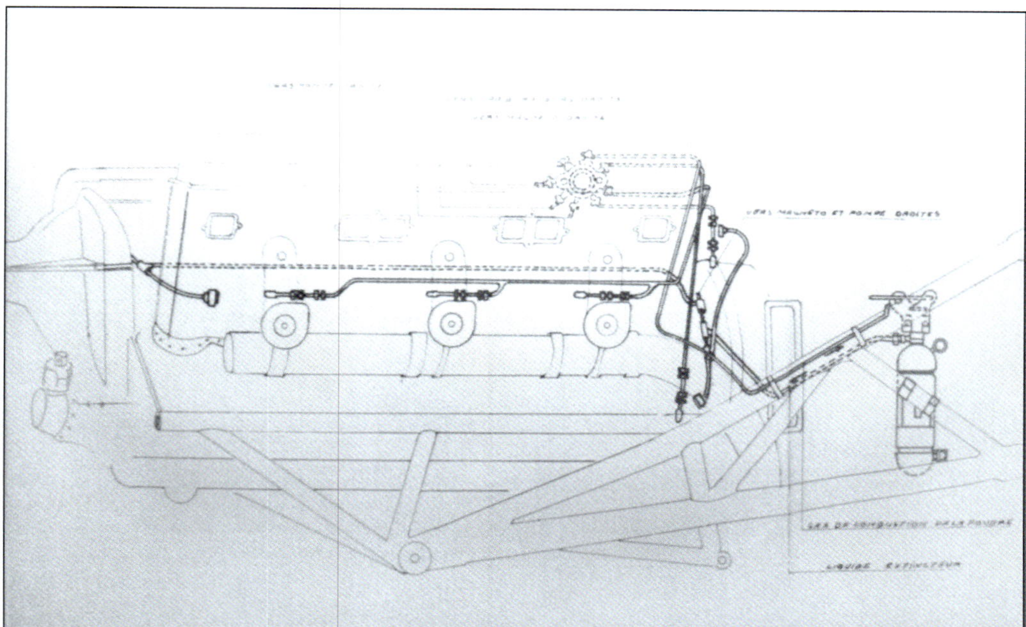

UNDERCARRIAGE

Right:
Main undercarriage, view from the right.
M. Lowe

Bottom:
Starboard main undercarriage leg, inner view.
B. Belcarz

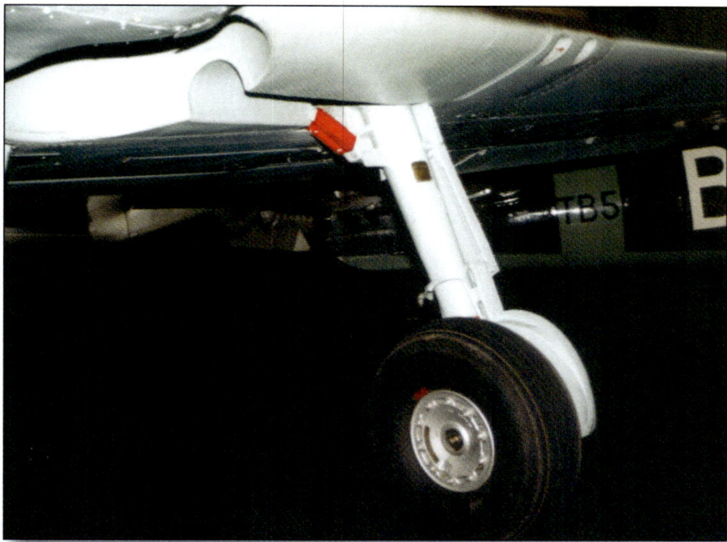

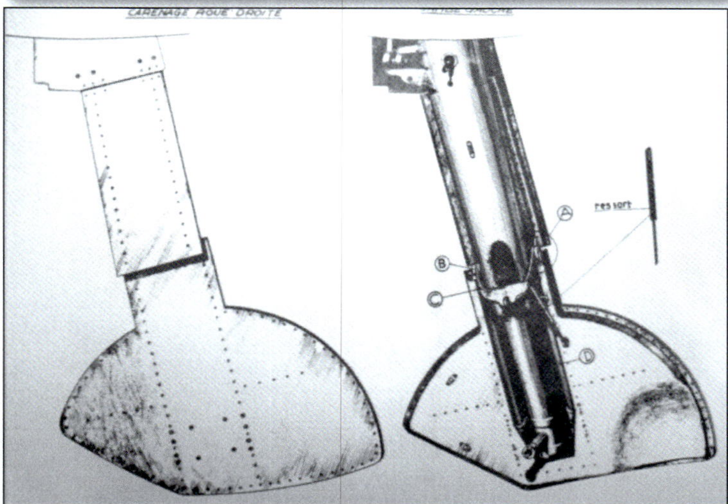

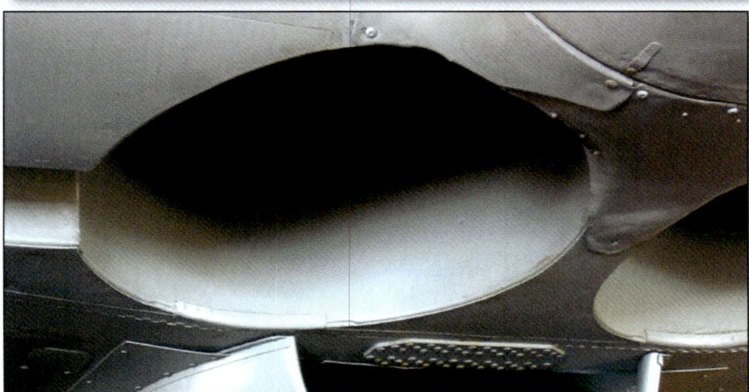

***Top**:*
Port main undercarriage leg.
B. Belcarz

Middle:
Main undercarriage leg covers, drawing from the Technical Manual

***Bottom**:*
Main undercarriage wheel wells
J. Kightly

Right:
The next photo of the main undercarrige wheel wells.
J. Kightly

Bottom:
Details of starboard main undercarrige leg and wheel.
J. Kightly

Two drawings of the main undercarriage details. Technical Manual

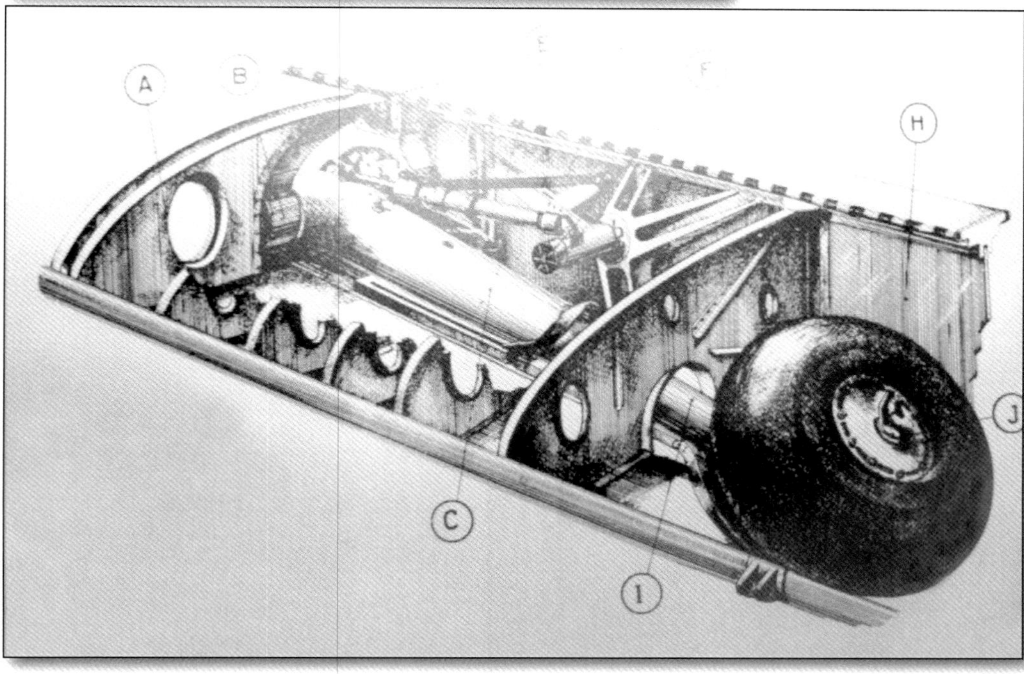

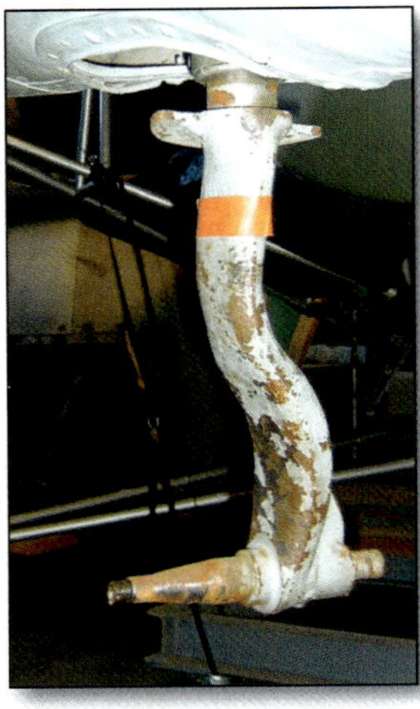

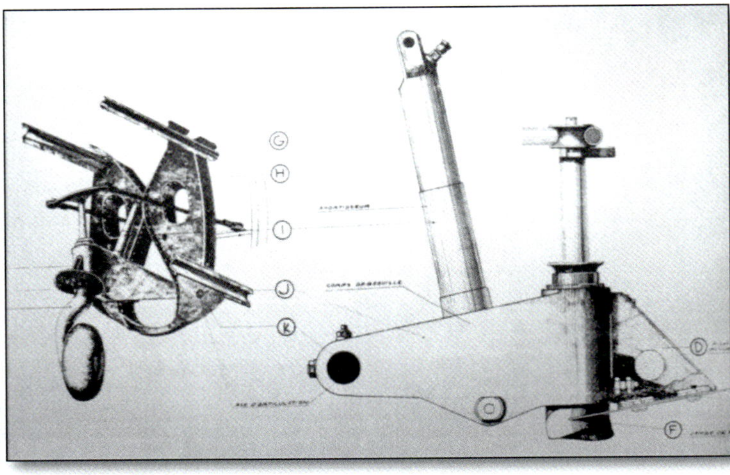

Above:
Tail wheel leg.
J. Kightly

Above right:
Details of the tail wheel, drawing from the Technical Manual.

Right:
Main undercarriage covers.
J. Kightly

Below:
Main undercarriage wheel – inner view
J. Kightly

D.520 "2", No. 421, S/Lt. Yves Rupied, 5 Esc. GC III/7, Toulouse-Francazal, June 1940.
Vert foncé/Brun foncé/Gris-bleu foncé – uppersurfaces with Gris-bleu clair undersurfaces.

D.520 "10", No. 73, Adj André Carrier, 2 Esc. GC I/3, Wez, May 1940.
Vert foncé/Brun foncé/Gris-bleu foncé – uppersurfaces with Gris-bleu clair undersurfaces.

D.520 "6", A/C Marie Emile Leblanc, 5 Esc. GC III/5, Relizanne, June 1940.
Vert foncé/Brun foncé/Gris-bleu foncé – uppersurfaces with Gris-bleu clair undersurfaces.

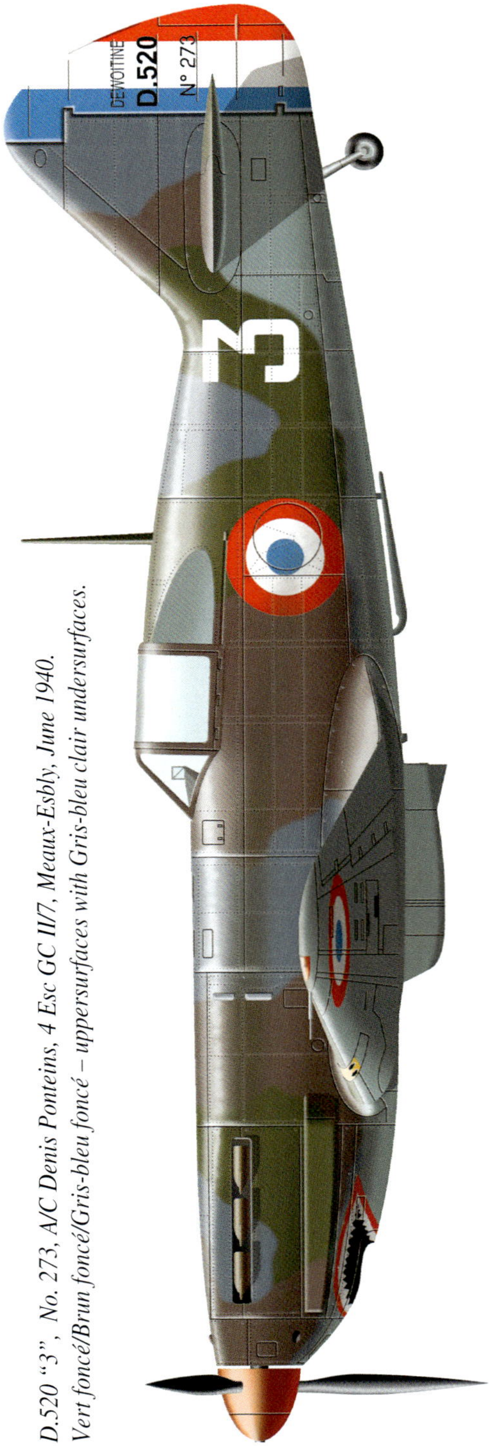

D.520 "3", No. 273, A/C Denis Ponteins, 4 Esc GC II/7, Meaux-Esbly, June 1940.
Vert foncé/Brun foncé/Gris-bleu foncé – uppersurfaces with Gris-bleu clair undersurfaces.

D.520 GC III/7, Toulouse June 1940. Vert foncé/Brun foncé/Gris-bleu foncé – uppersurfaces with Gris-bleu clair undersurfaces.

D.520 "30", A/C Bouton, 2 Esc. GC II/3, June 1940. Vert foncé/Brun foncé/Gris-bleu foncé – uppersurfaces with Gris-bleu clair undersurfaces.

D.520 "6", No. 266, S/Lt. Renè Pomier-Layrargues, 4 Esc. GC II/7.
Vert foncé/Brun foncé/Gris-bleu foncé – uppersurfaces with Gris-bleu clair undersurfaces.

D.520 "H", 3 Esc. GC II/7, Sgt. Henri Grimaud, June 1940. Usually pilots from this Escadrille used planes marked with first letter of their surname. Because in 3 Esc. there were two pilots with names starting with the letter "G", on Grimaud's plane (No. 230 or 104) there was the initial of his first name.

D.520 "M", No. 119, Mjr Mieczysław Mümler (Polish), 3 Esc. GC II/7,Ounans airfield, 15 June 1940. Vert foncé/Brun foncé/Gris-bleu foncé – uppersurfaces with Gris-bleu clair undersurfaces.

D.520 "11", Esc. AC2 Chasse de l'Aéronautique Navale (Navy Fighter Squadron) Hyères, 24 June 1940. Vert foncé/Brun foncé/Gris-bleu foncé – uppersurfaces with Gris-blue clair undersurfaces

D.520 "V", No. 242, S/Lt Georges Valentin, 3 Esc. GC II/7,Ounans airfield, 15 June 1940.
Vert foncé/Brun foncé/Gris-bleu foncé – uppersurfaces with Gris-bleu clair undersurfaces.

D.520 "I", No. 105, Cne. Bernard Challe, 2 esc. GC I/3, Oran-La Senia, August 1940. Vert foncé/Brun foncé/Gris-bleu foncé – uppersurfaces with Gris-bleu clair undersurfaces.

D.520 "GG" No. 347, 4 Esc. GC II/7, S/Lt Gabriel Gauthier, Sidi-Ahmed, Tunesia, September 1940. Vert foncé/Brun foncé/Gris-bleu foncé – uppersurfaces with Gris-blue clair undersurfaces

D.520 "11", Esc 1AC Chasse de l'Aéronautique Navale (Navy Fighter Squadron), North Africa 1940. Vert foncé/Brun foncé/Gris-bleu foncé – uppersurfaces with Gris-bleu clair undersurfaces.

D.520 "3", No. 38, Lt. René Lété, 3 esc. GC II/3, Homs, Syria 26 June1941. Vert foncé/ Brun foncé/Gris-bleu foncé – uppersurfaces with Gris-bleu clair undersurfaces. Tail Jaune Vichy.

D.520 "GG" No. 347, 4 Esc. GC II/7, Lt. Lt Gabriel Gauthier, Sidi-Ahmed, Tunesia December 1941. Proposed, but never used markings for Vichy aircraft. Vert foncé/Brun foncé/Gris-bleu foncé – uppersurfaces with Gris-bleu clair undersurfaces. Tail Rouge. Engine Rouge/Jaune Vichy

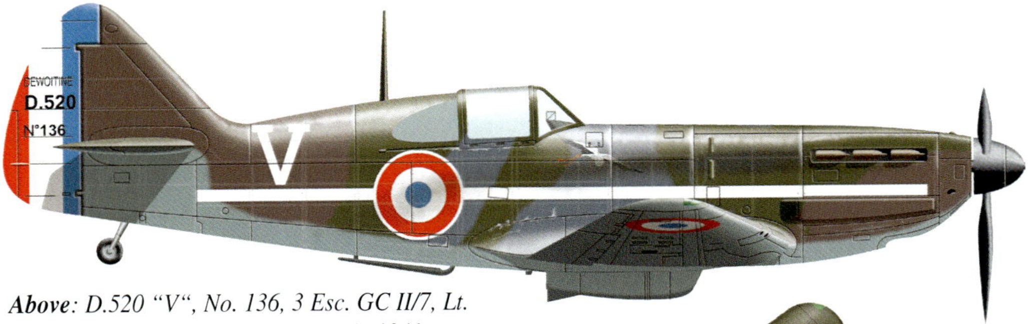

Above: *D.520 "V", No. 136, 3 Esc. GC II/7, Lt. Georges Valentin, Sidi-Ahmed, Early 1941. Vert foncé/Brun foncé/Gris-bleu foncé – uppersurfaces with Gris-bleu clair undersurfaces.*

Left and below: *D.520 "V", No. 136, 3 Esc. GC II/7, Lt. Georges Valentin, Sidi-Ahmed, 1942. Vert foncé/Brun foncé/Gris-bleu foncé – uppersurfaces with Gris-bleu clair undersurfaces. Tail and engine Rouge/Jaune Vichy*

D.520 "1", No. 350, 4 Esc. GC II/7, Lt. Marcel Merle, Sidi -Ahmed, Tunesia , end of 1941. Vert foncé/Brun foncé/Gris-bleu foncé – uppersurfaces with Gris-bleu clair undersurfaces. Tail and engine Rouge/Jaune Vichy.

D.520 "M", No. 173, Cdt. Jules Morlat, GC II/3, Alep-Nerab, July1941. Vert foncé/Brun foncé/Gris-bleu foncé – uppersurfaces with Gris-bleu clair undersurfaces. Tail Rouge/Jaune Vichy. Engine Jaune Vichy.

D.520 "M", No. 343, Cdt. Jules Morlat, GC II/3, Sfax, end of 1941. Vert foncé/Brun foncé/ Gris-bleu foncé – uppersurfaces with Gris-bleu clair undersurfaces. Tail Jaune Vichy.

D.520 No. 245, 3 Esc. GC II/7, Sgt. René Martin, Sidi-Ahmed, 1941. Vert foncé/Brun foncé/ Gris-bleu foncé – uppersurfaces with Gris-bleu clair undersurfaces. Tail and engine Rouge/ Jaune Vichy.

D.520 "1", 3 Esc. GC II/5, Lt. Pierre Villacèque, Casablanca 8 November 1942. Vert foncé/Brun foncé/Gris-bleu foncé – uppersurfaces with Gris-bleu clair undersurfaces. Tail and engine Rouge/Jaune Vichy.

D.520 "5", No. 55, Esc. 2AC, Second maitre Raymond Bedard, Port Lyautey, 8 November 1942. Vert foncé/Brun foncé/Gris-bleu foncé – uppersurfaces with Gris-bleu clair undersurfaces. Tail and engine Rouge/Jaune Vichy.

D.520 "12", Esc. 1AC (previously 1F), end of 1942. Vert foncé/Brun foncé/Gris-bleu foncé – uppersurfaces with Gris-bleu clair undersurfaces. Tail Jaune Vichy.

D.520 "1", No. 548, Reserve Captain Cliguet, Doret Group, Tarbes-Ossun, September 1944. RLM 75/ RLM 76 mix with small blotches RLM 74 – uppersurfaces with RLM 76 undersurfaces and bottom of fuselage.

D.520 "1", GC II/18 "Saintogne", France, Winter 1944/45. RLM 74/RLM 75 with small blotches in RLM 74 – uppersurfaces with RLM 76 Fuselage and undersurfaces

D.520 "37", GC Doret, Autumn 1944, just before all D.520s moved to GC II/18 "Saintogne". RLM 74/RLM 75 with small blotches in RLM 74 – uppersurfaces with RLM 76 Fuselage and undersurfaces.

D.520 No. 139, Douala airport, Cameroun 1940.
Vert foncé/Brun foncé/Gris-bleu foncé – uppersurfaces with Gris-bleu clair undersurfaces.

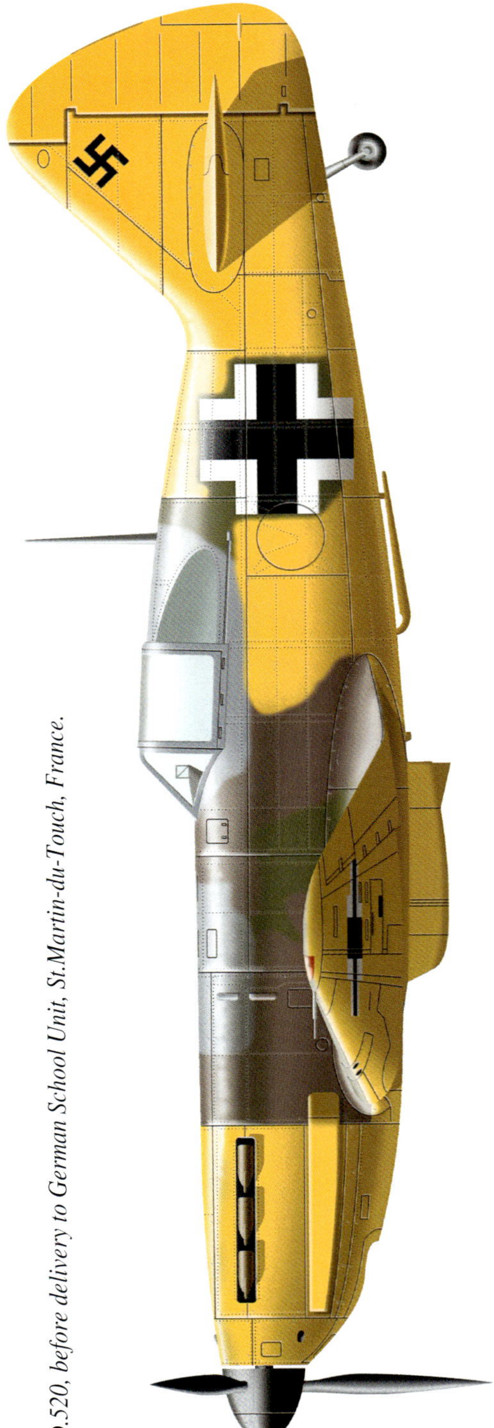

D.520, before delivery to German School Unit, St.Martin-du-Touch, France.

D.520 "NE+HZ" vel "94", JG 101 Luftwaffe, Pau France 1943. RLM 74/RLM 75 with small blotches of RLM 74 – uppersurfaces with RLM 76 fuselage and RLM 04 tail and undersurfaces.

D.520 "ON+FV" vel "84", JG 101 Luftwaffe, Pau-Pont-Long France, early 1944. RLM 74/RLM 75 with small blotches of RLM 74 – uppersurfaces with RLM 76 fuselage and RLM 04 undersurfaces.

D.520 "SV+GB", No. 544, JG 103 Luftwaffe, Brussels-Evere airfield, September 1944.
RLM 74 uppersurfaces, fuselage RLM 76, RLM 04 undersurfaces.

D.520 "85", JG 101 Luftwaffe, Pau-Pont-Long France, early 1944.
RLM 74/RLM 75 with small blotches in RLM 74 –uppersurfaces with RLM 76 fuselage and RLM 04 undersurfaces.

D.520, before delivery to German School Unit, St.Martin-du-Touch, France.
Vert foncé/Brun foncé/Gris-bleu foncé – uppersurfaces with RLM 04 undersurfaces, engine and tail.

D.520 "05+60", Factory airfield Toulouse-Francazal, 1943/44.
Vert foncé/Brun foncé/Gris-bleu foncé – uppersurfaces with RLM 04 undersurfaces, engine and tail.

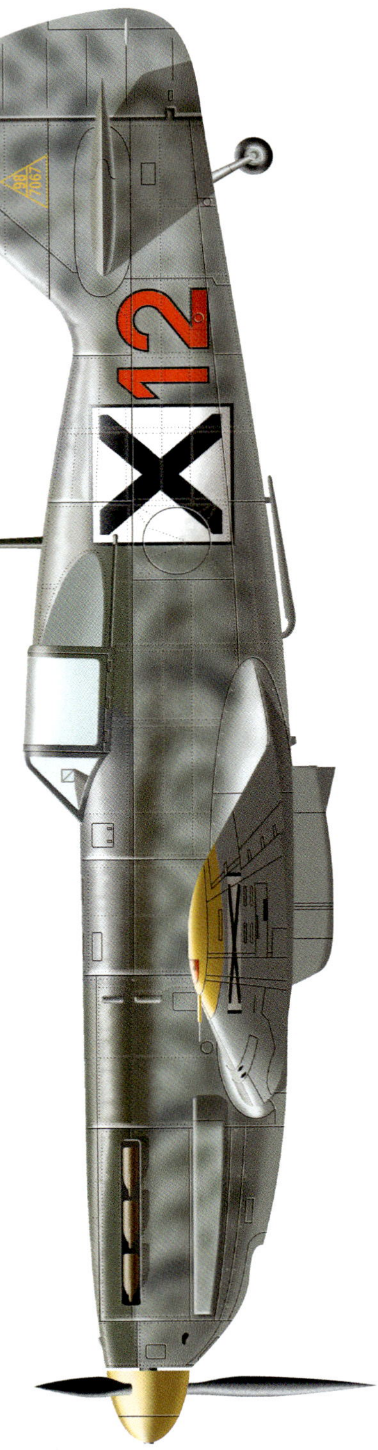

D.520 "12", podporuchnik (Pilot Officer) Stefan Marinopolski 652 Jato (Squadron), 6 Istrebitelnyj Orliak (Fihgter Group), Vrashdebna airfield near Karlovo, 1943. RLM 74/RLM 75 with small blotches in RLM 74 – uppersurfaces with RLM 76 fuselage and undersurfaces.

D.520, Polkovnik (Group Captain) Wassil Wolkow, Commander of 6 Istrebitelnyj Orliak (Fihgter Group), Marino Pole airfield near Karlovo, 1944. RLM 74/RLM 75 with small blotches in RLM 74 – uppersurfaces with RLM 76 fuselage and undersurfaces.

D.520 "14", 6 Istrebitelnyj Orliak (Fighter Group), Marino Pole airfield close to Karlovo, 1943. RLM 74/RLM 75 with small blotches of RLM 74 – uppersurfaces with RLM 76 fuselage and undersurfaces.

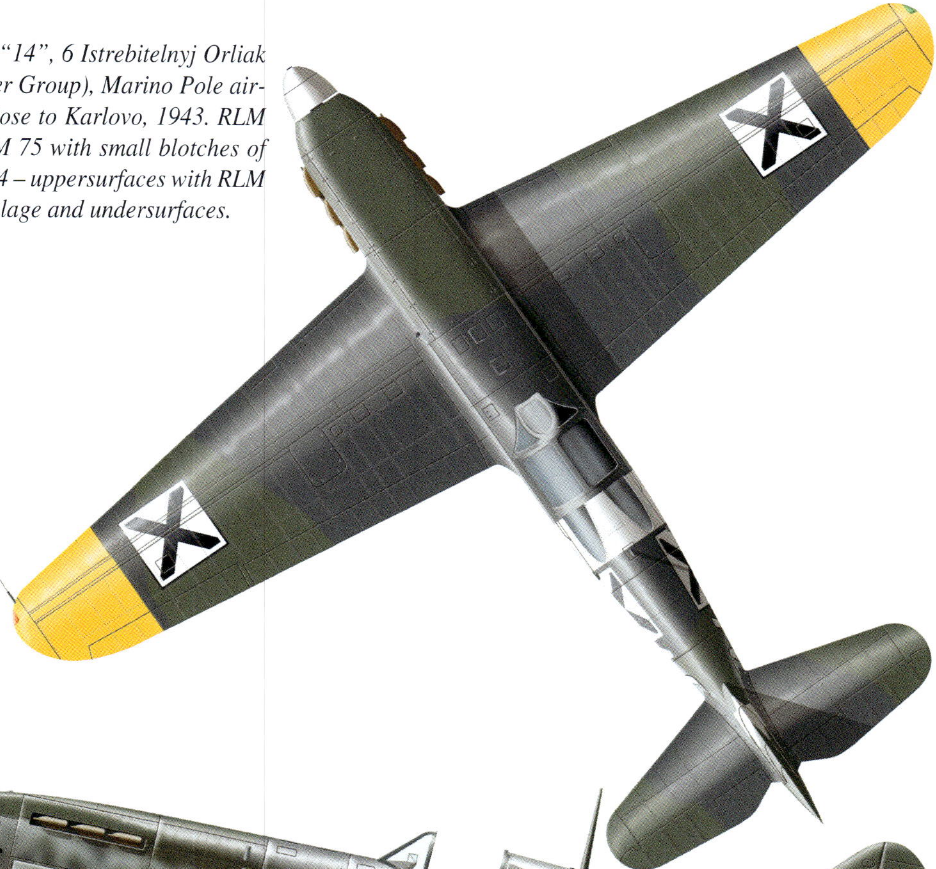

D.520 "13", 6 Istrebitelnyj Orliak (Fighter Group), Vrashdebna airfield near Karlovo, 1943. RLM 74/RLM 75 with small blotches in RLM 74 – uppersurfaces with RLM 76 fuselage and undersurfaces.

D.520 "12", Lt. Frixa, 59° Gruppo, 232° Squadriglia, Metato (Pise), Italy 1943. Nocciola Chiaro 4 with small blotches in Verde Oliva Scuro 2 – uppersurfaces with Grigio Azzuro Chiaro – undersurfaces.

D.520 “2”, Regia Aeronautica, Isteres, France, February1943. Vert foncé/Brun foncé/Gris-bleu foncé – uppersurfaces with Gris-bleu clair undersurfaces.

D.520 “14”, Regia Aeronautica, Isteres, France, February1943. Vert foncé/Brun foncé/Gris-bleu foncé –uppersurfaces with Gris-bleu clair undersurfaces

D.520 "20", Regia Aeronautica, Isteres, France, March 1943. Vert foncé/Brun foncé/Gris-bleu foncé – uppersurfaces with Gris-bleu clair undersurfaces.

D.520 "2", probably 101° Gruppo Autonoma Caccia, Aeronautica Nazionale Repubblicana, Mirafiori, January 1944. Vert foncé/Brun foncé/Gris-bleu foncé – uppersurfaces with Gris-bleu clair undersurfaces.